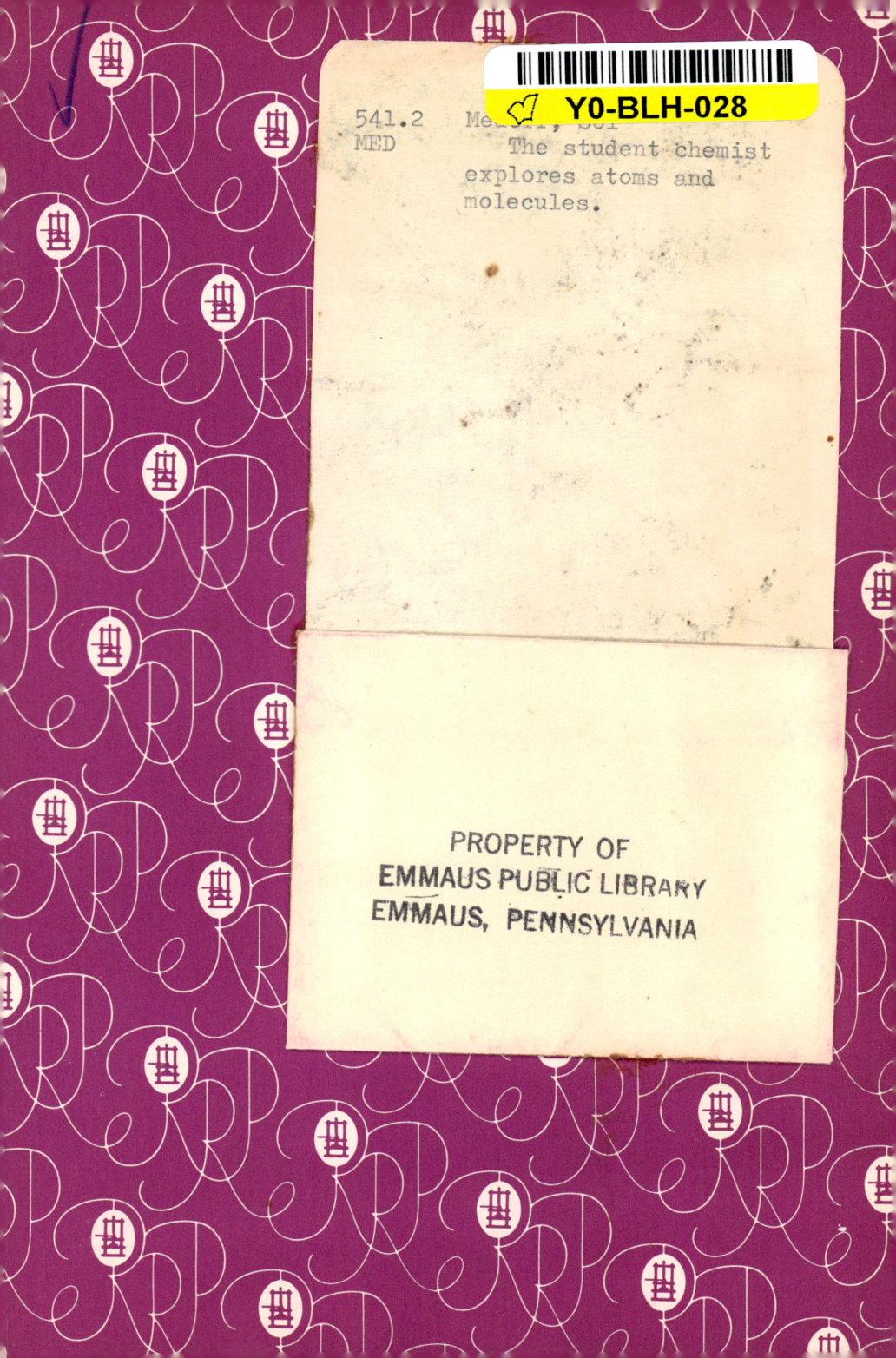

541.2 Medill, Sol
MED   The student chemist
         explores atoms and
         molecules.

PROPERTY OF
EMMAUS PUBLIC LIBRARY
EMMAUS, PENNSYLVANIA

THE STUDENT CHEMIST
EXPLORES ATOMS AND MOLECULES

THE STUDENT CHEMIST EXPLORES ATOMS AND MOLECULES

By

Sol Medoff and *John Powers*

Illustrated by
Nancy Lou Gahan

RICHARDS ROSEN PRESS, INC.
New York, N.Y. 10010

Published in 1977 by Richards Rosen Press, Inc.
29 East 21st Street, New York, N.Y. 10010

Copyright 1977 by Sol Medoff and John Powers

All rights reserved. No part of this book may be reproduced in any form without written permission from the publisher, except by a reviewer.

FIRST EDITION

Library of Congress Cataloging in Publication Data

Medoff, Sol.
   The student chemist explores atoms and molecules.

  (The Student scientist series)
  Bibliography: p.
  1. Atomic theory—Juvenile literature. 2. Molecular theory—Juvenile literature. I. Powers, John C., joint author. II. Title.
QD461.M43   541.2   76–51386
ISBN 0–8239–0381–8

Manufactured in the United States of America

About the Authors

Sol Medoff is a product of the New York City University system. Born of immigrant parents, he early found schooling to be a secondary consideration compared to "making a living." At the age of twelve he began to work to help support his family.

During World War II he served in the U.S. Army, part of the time in charge of a chemistry laboratory in New Caledonia. After the war he and a friend opened a textile-dyeing business in New York City.

Twelve years later he received a full scholarship from New York State, with a personal letter from then Governor Averell Harriman, to complete his doctoral degree. In 1956 he began to teach at Brooklyn Technical High School. Currently he is teaching at Westbury High School in Westbury, Long Island. For a number of years he was chairman of the North Nassau branch of the Science Teachers Association of New York. In 1970 he introduced the formal study of biochemistry on a high-school level; he teaches the subject to selected students.

Mr. Medoff is married and has two children, a boy and a girl.

John Powers has had a lifelong interest in things scientific. From early childhood in Montana this fascination took him to Yale University for a B.S. degree in chemistry and to Harvard University for a Ph.D. in organic chemistry. Upon leaving Harvard he was employed in the scientific industry, studying such diverse fields as natural products chemistry, synthesis under high pressure, photoconductivity, and electro-optic effects.

He began his academic career at Hunter College of the City University of New York and assumed his present post at Pace University in 1970. He is the author of a textbook on organic chemistry and many scientific papers. His main interest now centers on the detailed study of biochemical reactions.

Mr. Powers lives in Floral Park, New York, with his wife, two daughters, an assorted menagerie, and a large collection of old phonograph records, all of which occupy his spare time.

Contents

I.	The Awakening of Chemistry	3
II.	Opening Atomic Secrets	12
III.	Setting the Chemical House in Order	21
IV.	The Marriage of Elements	35
V.	How Strong Is a Chemical Wedding?	50
VI.	The Remarkable Aspects of Carbon and Silicon	57
VII.	How Small Molecules Grow into Giants	67
VIII.	The Power Locked in the Nucleus	78
IX.	Let's Do Some Molecular Research	88
X.	How to Win the Nobel Prize in Chemistry	96
	APPENDICES	
	I. Glossary	104
	II. Books for Further Reading	115

List of Illustrations

Fig.
1. Making iron with magic
2. A few of the symbols used by Dalton
3. Electron beam tube with magnetic and electric deflectors
4. Thomson's picture of an atom
5. Rutherford's experiment
6. The Rutherford atom
7. The type of pictures that were drawn by alchemists
8. Part of Beguyer's Telluric Screw
9. Part of Newlands' table
10a. The Periodic Table of Mendeleev
10b. The Periodic Table as used today in chemistry
11. Combinations of atoms
12a. Electrostatic attraction
12b. Sodium chloride crystal
13. Methods of bonding
14. Structure of water
15. Water molecule with bonding value
16. Polar nature of water molecule
17. Hydrogen bond
18. Methane molecule
19. Induction effect
20. Coordinate covalent bond
21. Covalent bond with hydrogen ion
22. Potassium chlorate
23. Testing strength of bond
24. NaCl under attack in water

LIST OF ILLUSTRATIONS

25. *Row 2 of Periodic Table*
26. *Carbon atom*
27. *Geometry of carbon compounds*
28. *Structure of glucose*
29. *Double bonds*
30. *Structure of benzene*
31. *Giant molecules*
32. *Two types of sulfur cross-links*
33. *A peptide link*
34. *Structure of nylon*
35. *Changing of uranium to lead*
36. *Isotopes of uranium and lead*
37. *Entries in research notebook*

Preface

Some time ago a young woman visited me in my laboratory. Although the call was purely a social one, she could not refrain, out of curiosity, from asking me a question. Staring at the apparatus on the lab table, she said, "Why does the water stay in the bottle, and not fall down?"

She was looking at some receiving bottles, all completely filled with water. They rested upside down on shelves in a trough. The shelves had holes, and the bottle mouths were over these holes. Water from the trough came up over the bottle necks. I was all set to collect some hydrogen gas. This method of displacing water with hydrogen is a good one, because hydrogen is virtually insoluble in water.

My answer to the young woman's question was short and simple. I merely pointed out that the air pressure on the water in the trough was strong enough to hold the "bottle-water" up.

The incident set me thinking, however. I wanted to give a much clearer picture of air molecules. In fact, I wanted to give a mental picture of atoms and molecules in general. I found the idea exciting and, together with my friend John Powers, began to write this book. It has been a great pleasure working on it, and we hope you find pleasure in reading it.

SOL MEDOFF

THE STUDENT CHEMIST
EXPLORES ATOMS AND MOLECULES

I

The Awakening of Chemistry

About two thousand years ago Greek philosophers began to discuss the possible existence of tiny, indivisible particles. The Greek word for indivisible is "atomos," and thus the "atomic theory" was born. People would not believe such an abstract idea, however, and the atomic theory was nearly lost for many centuries.

In those early days, there was no real experimentation. The so-called scientists had long complicated "recipes" for making medicines and for changing one metal into another. Some of these scientists, or alchemists as they were called then, became famous; even kings employed them. An Arab, Abu Musa Jabir, wrote in such mystic and symbolic language that his recipes were difficult to decipher. The famous Raimundus Lullus of Spain said that alchemy was a talent that could not be learned or taught. Yet Lullus did give long and complicated recipes for making gold from base metals and for concocting medicines.

Now let us imagine we are in a tribal town in Africa. A large furnace is constructed in a clearing in the jungle, away from the houses of the town. Special kinds of stones are gathered and placed in the furnace. With the stones are placed many logs of wood. A fire is started. It is a great and sacred fire, for the witch doctor has come to perform rituals. He calls upon the gods to change the stone into useful metal. After the fire burns well and charcoal is produced, the iron ore is indeed reduced to iron. The iron will be used to make tools, jewelry, and spears. The ore has always been changed to iron through the witch doctor's magic. Part of the recipe for the production of iron was the hocus pocus and dancing and waving of arms. This was "chemistry."

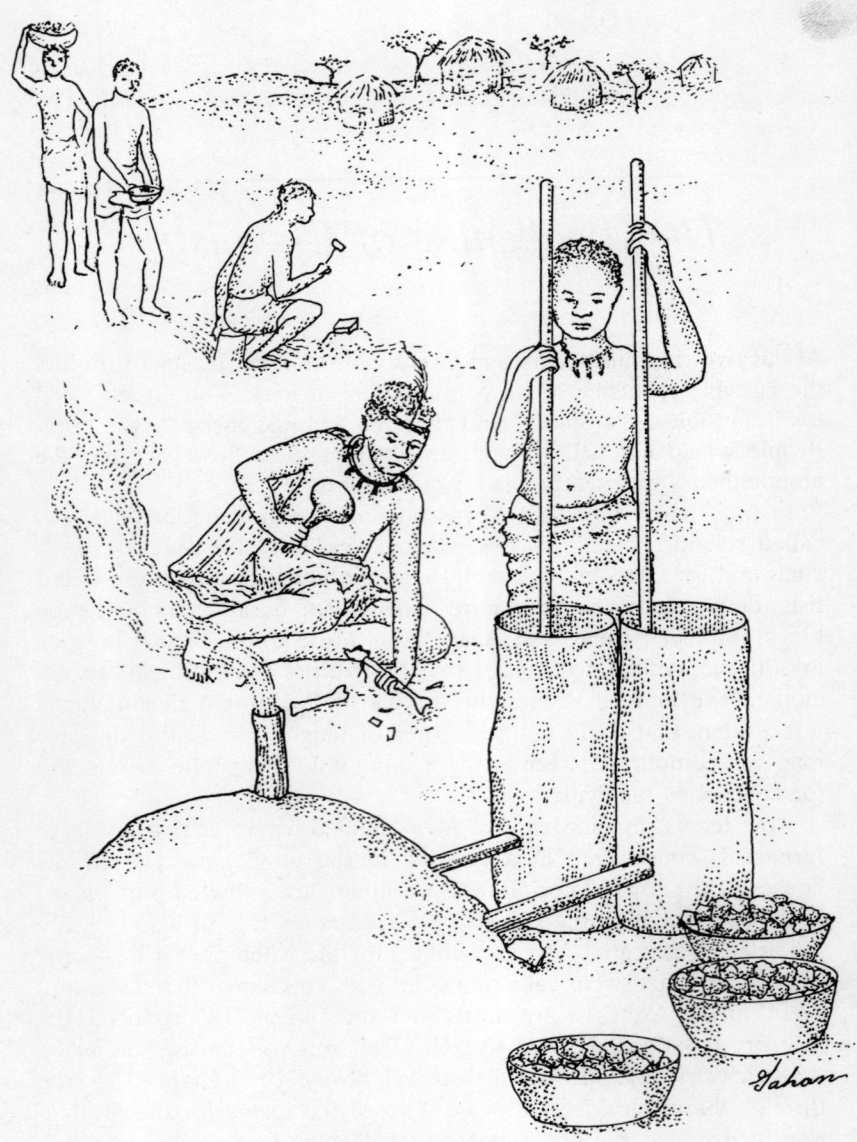

FIG. 1. *Making iron with magic*

The Phlogiston Theory

In order for us to understand the recent nature of the fantastic science of chemistry, let us examine a theory that arose in 1702. Two German scientists, Johann J. Becher and Georg E. Stahl, proposed what they called the "phlogiston theory." This theory was believed until 1789, by which time the industrial revolution was in full bloom, the United States had achieved independence, and Benjamin Franklin was investigating electricity.

The theory stated that all combustible material contains an element called phlogiston. When the material burns, phlogiston escapes. According to the theory, all metals, when heated, lose phlogiston and become a substance called "calces." Because charcoal contains a lot of phlogiston, heating calces with charcoal causes the metal to regain phlogiston and become a true metal again.

What really happens, as we know today, is that a metal, on being heated, combines with oxygen. Thus charcoal, when heated with the metal oxide, combines with the oxygen, leaving the metal free once again:

$$\text{Iron} + \text{oxygen} = \text{Iron oxide}$$
$$\text{Carbon (charcoal)} + \text{iron oxide} = \text{Carbon oxide} + \text{iron}$$

Would it not have been a simple matter to have weighed things before and after a reaction? The amazing fact is that few, if any, quantitative experiments were done at that time. Weighing, we suppose, was only useful for selling fish and meat. It is interesting that when hydrogen gas was discovered, it was thought to be phlogiston!

In every age, genius rears its head. In 1500 Leonardo da Vinci said that air had to contain at least two ingredients, because some of it was left after burning. No one then appreciated that statement.

We come finally to the destruction of the phlogiston theory. The advancement of science had stumbling blocks, and this theory was one of them. The beginning of its end was attributed to an English clergyman, Joseph Priestley (1733–1804). Priestley wrote a book entitled *Experiments and Observations on Different Kinds of Air*. His home was near a brewery, and he must have been friendly with the owner. We know that he collected carbon dioxide—the bubbles in beer—and passed the gas into pure water. He liked the taste of the mixture and made a lot of it. He gave it to all his friends, and soda water was soon very popular. Priestley is credited with the discovery of oxygen, although it was also identified independently by the Swedish chemist Karl W. Scheele. Scheele is largely forgotten, but in his short life (he died at forty-four) he also discovered the elements chlorine and manganese.

Although Priestley had much to do with the destruction of the phlogiston theory, he himself believed in it as long as he lived. His good friend from America, Benjamin Franklin, was engaged at the time in a deep study of electricity. He also believed in the phlogiston theory.

Priestley was a Unitarian minister in Leeds and Birmingham, in northern England. He published a great deal of material and was respected as a good experimenter. But not all of his publications were on chemistry. He went deep into politics and was an extreme liberal. He sided strongly with the revolutionary forces in France. In 1791, on the second anniversary of the French Bastille Day, Birmingham was actually under siege by conservatives, who hated liberals. They burned Priestley's church, his home, his laboratory, and his manuscripts. Priestley spent the next three years in London, and then sailed to America and his friend Benjamin Franklin.

Priestley had a friend in France who was an excellent chemist. They were quite different, for Priestley was openly communicative whereas his friend, Antoine Laurent Lavoisier (1743–94), was quite secretive. Priestly told Lavoisier all about his experiments with oxygen, and Lavoisier quietly put together a series of experiments.

Before we go further with the experiments of Lavoisier, we must mention another chemist, Robert Boyle (1627–91), who had a direct influence on the Lavoisier experiments with oxygen. Boyle actually weighed things! He was greatly annoyed with alchemists. He wrote, "Finding the generality of those addicted to chemistry to have had scarce any view but to the preparation of medicines, or the improvement of metals, I was tempted to consider the art not as a physician or alchemist, but as a philosopher." Boyle weighed air; he weighed it compressed, natural, and expanded. He almost discovered oxygen, as his experiments involved a great deal of burning. Boyle wrote, "Some part of air is necessary for both combustion and respiration."

By stating that air was composed of several kinds of tiny particles, he reopened the atomic theory. Today Boyle is called by many the "father of chemistry." Lavoisier, influenced by Boyle's experiments, is known as the "father of *modern* chemistry."

The stage was set for Lavoisier by Boyle with his methods of weighing things, and by Scheele and Priestley with their discovery of oxygen. About the phlogiston theory Lavoisier wrote, "That substance [phlogiston] is sometimes endowed with levity, and at other times it has weight; sometimes it can penetrate a brick, and sometimes it cannot. It has all kinds of wonderful attributes, to which I shall add another one; I am going to show it is just a figment of the imagination."

Conservation of Matter

Lavoisier discovered that when he heated "mercury calx" (a mercury-oxygen compound), it changed to mercury metal. At this point, Lavoisier measured the amount of oxygen released. He succeeded in demonstrating that the same amount of oxygen was consumed by the mercury to form the mercury-oxygen compound. He then wrote, "In all the operations of art and nature, nothing is created; an equal quantity of matter exists before and after the experiment—upon this principle, the whole art of performing chemical experiments depends." Lavoisier's principle is known today as the "Law of Conservation of Matter."

Lavoisier was not only a fine chemist; he was also a tax collector. As a tax collector, he worked for an agency that did business with the Royal French Government. When the revolution came, Lavoisier was remembered for his association with the aristocrats, and in 1794 he was hailed before the tribunal. His plea for clemency was rejected. Coffinhal, president of the tribunal, said, "The republic has no need for chemists and savants. The course of justice shall not be interrupted." Lavoisier was beheaded.

It was in 1789 that Lavoisier had published his now famous textbook, *Fundamental Treatise of Chemistry,* which was the turning point for chemistry as a science. Besides upsetting the phlogiston theory and establishing the principle of the conservation of mass, it began a system of naming elements and compounds that essentially is in use today.

In all this time, with great experiments being performed, what happened to the atomic theory? Perhaps the truth is that there didn't seem to be a need for the theory. The pharmacist had no use for it, nor did the physician. Industry produced textiles and bricks without an atomic theory. Still, the idea would not perish of neglect. The Greeks who first talked about indivisible particles seemed to cry with the wind that they would be heard.

Atomic Theory

An English chemist, John Dalton (1766–1844), was listening. In 1803 he wrote about the atomic theory. Again, in 1808, he expanded on the same theme. Most of the world didn't care much one way or the other. Many, in fact, did not believe it. Although Dalton reawakened the entire atomic concept in the beginning of the 19th century, it was not until almost the beginning of the 20th century that the theory gained real substance. Dalton's points were:

a. All matter is composed of small, indestructible particles.
b. Atoms of the same element are alike in mass.
c. Atoms of different elements are different.
d. Reactions involve the entire atom, never part.
e. Elements combine in ratios of small whole numbers.

Dalton was wrong in saying that atoms of the same element are all alike in mass. In fact, Dalton made many errors in his papers about the composition of matter; but his work was the springboard for future experimentation on the atom. Some of Dalton's symbols for the elements are illustrated in Fig. 2.

It is interesting to note that Dalton was colorblind. He was a Quaker, with a Quaker's habit of dressing in somber colors, very simply, and carrying no weapons. One day the king of England, having heard of the great chemist, invited Dalton to the palace. Now, to appear before the king, it was necessary to wear colorful court clothes, with a beautiful sword hanging at one's side. Dalton did the only thing he could. He put on a scarlet robe in which he had received an honorary doctorate degree from the University of Oxford. The robe was so full that no one could be sure that a sword did not,

The Awakening of Chemistry

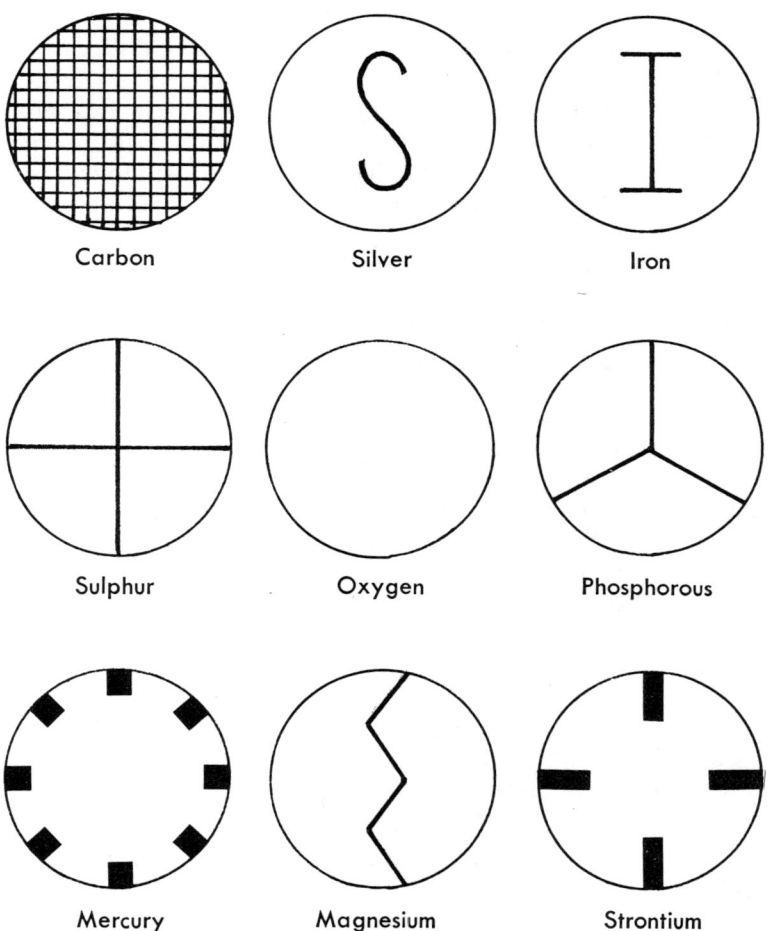

FIG. 2. *A few of the symbols used by Dalton*

indeed, swing underneath. Thus Dalton made his appearance before the king. Dalton's Quaker friends took him to task about the scarlet robe, saying that the color was against their religious principles. Dalton replied; "To thee it is scarlet, but not to me. It is green like the color of leaves."

Dalton extended the idea of air atoms (molecules) to atoms in liquids and solids. He tried to show by drawings the difference between liquid phases and solid phases.

Often history omits the names of persons who have had strong impact on scientific thinking. One such person was William Prout (1785–1850). Prout discovered that the acid in the stomach was none other than hydrochloric acid. He also wrote about an idea so strange and bizarre for his time that he would not sign his name to the publication, for fear of being ridiculed. He asked whether it was not possible that all elements were really compounded from one element, hydrogen. Hydrogen, being the lightest and therefore the simplest of all elements, could group to make all the other elements. In this case, taking the hydrogen atom as a mass of one, all atomic masses should be whole numbers. But wherever scientists had carefully calculated the masses of the elements, they had all turned out to be numbers with fractions. In fact, many scientists, taking Prout's publication quite seriously, tried to be extremely careful in their measurements. Yet, they always ended with fractions in the mass number. As we know today, the measurements were accurate, but Prout was also on the right track.

Invention of Electric Battery

Not until 1910 was Prout vindicated. Meanwhile, scientists began to get help from an unexpected source. Count Alessandro Volta (the word "volt" comes from his name) had built a current-generating battery in 1800. He did this by using plates of copper and plates of zinc, separated by moist cardboard. If one wire is attached to the copper and one to the zinc, an electric current is produced. The battery proved to be a tremendous boost to the discovery of new elements, and even to the discovery of atomic particles. One of the first experiments with electricity was the decomposition of water into hydrogen and oxygen gases.

Isolation of Elements

The scientist of the day most noted for the discovery of new elements was Sir Humphry Davy (1778–1829). Davy had been knighted on the day before his wedding. He had been an apothecary's apprentice, and had studied mathematics and science in his spare time. He also loved to translate Latin poems into English.

Knowing about the separation of water with the aid of electricity, Davy attempted to separate other elements from their compounds. Most elements are rarely seen in the pure state, but rather occur in combination with other elements. Elements differ in the ease of combining with other elements, and the most "active" elements form the strongest bonds. One such element is potassium. Davy separated potassium in its pure state with the new tool, electricity. Another very active element, closely related to potassium, is sodium. Davy separated sodium, as well. Indeed, a new tool had been discovered, and with Davy's experiments, it excited the entire scientific world. We could continue with many similar experiments by others (Berzelius, for example), but instead let us return to some famous experiments dealing with atomic particles.

II

Opening Atomic Secrets

In 1897 Joseph John Thomson (1856–1940) was experimenting with an electric beam. He used a tube from which the air had been evacuated. In such an environment the air resistance is negligible, and the electric particles could move easily. There are "minus" and "plus" poles in electricity, and they are called "electrodes." The electric beam traveled from one electrode at one end of the tube to another electrode at the other end of the tube. Thomson then subjected the beam to both electric and magnetic force fields outside of the tube. He noticed that the lightninglike electric beam was deflected in the presence of the fields, and he determined that the beam was composed of minus particles. He called them "electrons."

Thomson had been careful with his measurements. He related the amount of deflection of the beam to the strength of the magnetic field. In the mathematical equations that resulted, he discovered that it was also necessary to know the velocity of the beam. The velocity of the beam would be equivalent to the velocity of individual electrons. What Thomson wanted to find was the mass of the electron and the electrical charge on the electron.

To find the velocity, Thomson set up an experiment using the reasoning: "As the beam shot forward, what force would be necessary to just stop it?" The force that would change the forward motion to zero motion would be equivalent to the force that gave momentum to the electron beam. Momentum equals mass times velocity. Thomson set up a variable magnetic force that could be adjusted so that it just stopped the beam from moving. His experiments proved to be accurate, and his calculations, the work of a genius.

The velocity of the electron proved to be $\frac{1}{10}$ the speed of light. This answered a question that had bothered scientists. Electrons did not seem to be affected by the pull of gravity. Now, Thomson's explanation was that particles traveling at such high speeds were not visibly affected by the gravitational pull.

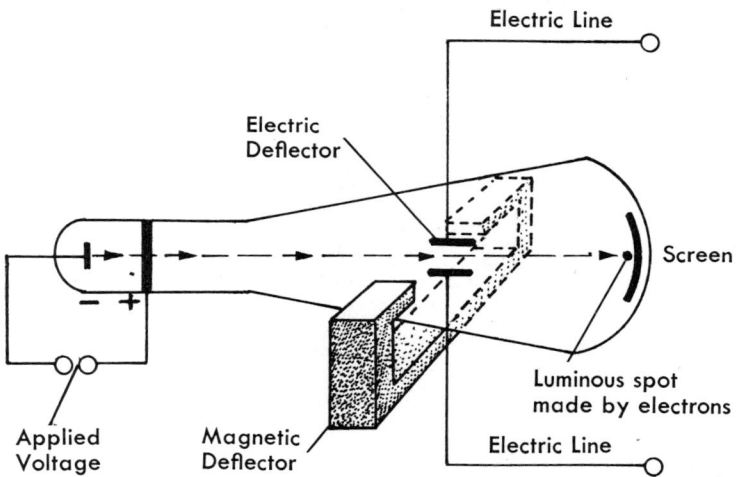

Fig. 3. *Electron beam tube with magnetic and electric deflectors*

Charge to Mass Ratio

Another result of Thomson's experiments was the discovery of the "charge to mass" ratio for the electron. It is one thing to say that a particle has a certain amount of charge, and still another to say how much of a mass that charge is spread over. If the same amount of charge is on a small mass, it is much more intense than if it were on a larger mass. Thomson had also studied the behavior of "positive rays," which act exactly opposite to the electron beam. If positive rays were generated from hydrogen gas, the charge to mass ratio could be obtained for a positively charged hydrogen particle (proton). This charge to mass ratio was $\frac{1}{1836}$ of the value of the charge to mass ratio of an electron. The hydrogen nucleus has an equal and opposite charge compared to the electron. The reasoning now ran that the neutral hydrogen atom must contain an electron and a nuclear particle called a proton. Calculations started by Thomson put the mass of an electron at 9.1091×10^{-28} gram today. Since the electron has been demonstrated to be of such small mass, the phenomenon of electricity running through wires can be appreciated.

Further, Thomson drew a picture of the atom as he envisioned it. The atom was now a sphere with a variable density. The inside,

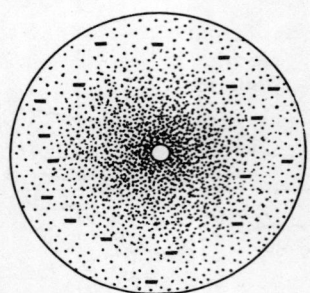

FIG. 4. *Thomson's picture of an atom*

starting with the nucleus, was the densest of all. The sphere became less and less dense toward the perimeter. The least dense portion contained the electrons.

Although the Thomson atom had variable density, it was still a solid sphere. Dalton had also pictured the atom as being solid, but not with the variable density and the electrons farthest from the nucleus.

A number of scientists did not agree with the Thomson picture. One of them was a New Zealander, Ernest Rutherford (1871–1937). Rutherford performed a series of experiments that rank with the greatest ever performed, and indeed, with those of Thomson.

Radioactive Elements

In his experiments, Rutherford used an element with an unstable nucleus. A number of elements have unstable nuclei because, it is

Opening Atomic Secrets 15

assumed, of the way the particles of the nucleus are arranged with respect to each other. These nuclei constantly decompose, most often changing to another element, but in any case going in the direction of stability. In decomposing, the nuclei shoot out pieces, much like bullets from a gun. Such elements are called "radioactive" elements and can be dangerous to living cells.

The type of radioactive element that Rutherford chose ejected relatively large particles, each with an atomic mass of 4 (equivalent to 4 hydrogen nuclei or 4 protons). The particle was known to be the helium nucleus, the element helium having a mass of 4. The charge on the particle was not 4+, however, but rather a puzzling 2+. It is to be remembered, at this point, that the particles used by Rutherford were tremendously more massive than the Thomson electrons.

Rutherford aimed his "machine gun" (trillions upon trillions of helium nuclei or "alpha" particles, as they are called, were shot out) at a sheet of gold leaf only $\frac{1}{50,000}$ centimeter thick. In terms of the number of atoms in such a tiny thickness, however, the alpha particles had to penetrate 20,000 atoms. This was like having a defense wall 20,000 bricks thick against the machine gun. Behind the gold

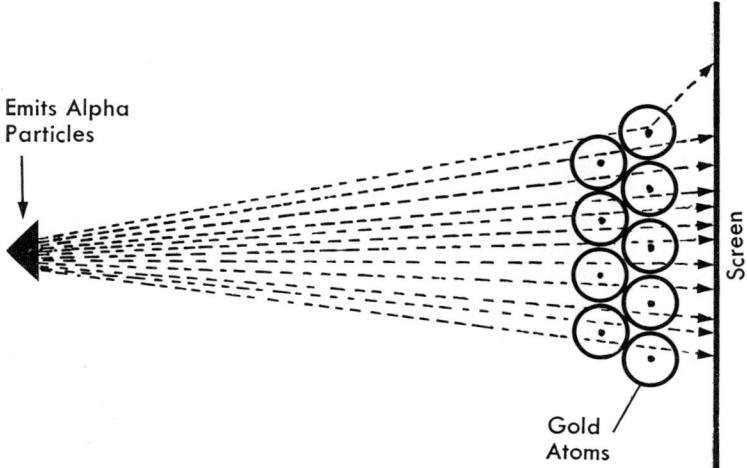

Fig. 5. *Rutherford's experiment*

leaf Rutherford placed a photographic plate, to detect atomic penetration. Later, he and his young workers were to place photographic plates in every conceivable position, except directly in front of the gold leaf in the path of the bullets. For example, to see if the alpha bullets ever bounced backward, he placed photographic plates behind his "machine gun."

Rutherford had begun his experiments to prove that the atom was not a solid sphere. He maintained that atoms contained space. But even he was completely amazed at the results of his experiments. Almost all of the alpha particles went straight through the gold leaf as if it weren't there. We must appreciate the fact that the alpha particle has a mass 7,350 times the mass of the electron. It could, if electrons got in its path, blow them away like mere nothings. The large gold nucleus was another matter. Since plus repels plus, an alpha particle would be deflected if it even came near the gold nucleus. In that case, it might go in one of several directions. The photographic plates were set to catch pictures of these deflections. But the fact was that only one bullet out of 8,000 was actually deflected, the rest going straight through.

Rutherford and his team did many calculations. They determined the atomic size. They estimated the size of the nucleus. They calculated the diameter of the atom to be, roughly, 10^{-8} cm, and the diameter of the nucleus, 10^{-13} cm. The diameter of the nucleus was thus 100,000 times smaller than the diameter of the atom as a whole. Then, calculating sizes according to volume, they found that the nucleus was only a trillionth (10^{-12}) of the volume of the atom. For the first time, a good detective had discovered that matter was mostly space.

The Rutherford picture of the atom was that of a hollow sphere, with a massive nucleus in the center and electrons scattered in the surrounding emptiness.

Often, a clever scientist is called upon to interpret the meaning of results obtained in an experiment. Without ever performing the experiment himself, the scientist can contribute greatly to the advancement of scientific knowledge. Thus it was in 1932, when an English physicist, James Chadwick (1891–1974), interpreted certain phenomena observed by others. The same type of alpha particles used by Rutherford had been used to bombard the element beryllium.

Opening Atomic Secrets 17

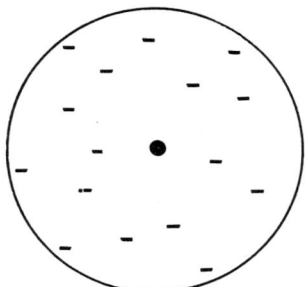

FIG. 6. *The Rutherford atom*

Strangely, because of the bombardment beryllium itself emitted a radiation, which was different from any ever seen before. It was a very penetrating radiation and was not affected by a magnetic field. This indicated, of course, that the radiation had no charge. When a substance such as paraffin was placed in the path of the radiation, protons were thrown out of the paraffin. What was happening?

Chadwick then published his interpretation, and all of the others agreed. Chadwick said that the mass of the radiation particle shot from beryllium must be about equal to the mass of the proton that was dislodged so easily from the paraffin. Electrons, being so light, could not behave in such a way. It was like the case of a billiard ball hitting an equally sized and equally massive billiard ball. Chadwick named the newly discovered particle "neutron," because of its neutral character, and said that it had about the same mass as a proton.

Nature of the Nucleus

It was the great German mathematician Werner Heisenberg (1901–   ), however, who first suggested that the neutron was part of the nucleus. The nucleus, he said, consisted of two main types of particles called "nucleons," the protons and the neutrons. Today we are aware of a much more complicated nucleus.

Consider, now, the nature of the nucleus with respect to the element it represents. A neutral atom contains the same number of electrons as do protons. If the nucleus contains two protons, the

18 THE STUDENT CHEMIST EXPLORES ATOMS AND MOLECULES

element is helium. It acts chemically like helium because of the two electrons external to the nucleus. It does not matter if the nucleus contains one, two, three, or more neutrons; but as long as there are simply two protons, the chemical nature of helium is set. In other words, atoms may differ in mass, but as long as they have the same number of protons, they belong to the same element. With the discovery of the proton and the neutron, it became possible to explain the fractional atomic weights. All that is necessary is that some nuclei of the same element have different numbers of neutrons. Nuclei of the same element that differ in mass are called "isotopes."

We now know that hydrogen has three forms: hydrogen of mass 1, of mass 2, and of mass 3. Hydrogen of mass 1 contains just one proton; of mass 2, a proton and a neutron; and of mass 3, a proton and two neutrons. Each isotope has a different name. The first is simply hydrogen. The second is deuterium; the third is tritium. The atomic mass of hydrogen is an average of the three and is 1.0078. Such a number indicates that the most common form of hydrogen is the isotope of mass 1. In fact, 99.98 percent of this element is mass 1, 0.02 percent is mass 2, and only a trace is mass 3. It is easy to separate hydrogen isotopes because they differ so much in weight. All that is necessary is to evaporate a great deal of water. The water molecules containing deuterium or tritium will lag behind, being heavier and slower. Finally, a water rich in these isotopes can be obtained. Such water is called "heavy water."

Rutherford's atom gave a better picture of the structure. But the picture was still too simple. The chemical nature of an element could not be understood without knowing how the electrons were actually arranged. Knowledge of arrangement would lead to understanding and predicting of reactions. Research could then go on to new and wonderful compounds.

Discovery of X Rays

In the early part of the 20th century, a useful tool was invented, a tool that helped reveal more about the atom. Thomson's "cathode ray" tube had been known for some years.

A scientist named Wilhelm Röntgen (1845–1923) noticed certain emanations from the glass of the Thomson tube. It was known that if charged particles (like electrons), going very fast, were suddenly slowed down, they would produce a magnetic field. Radiations would then be emitted. It was evident to Röntgen that the glass at the end of the tube was slowing the fast electrons; some form of radiation was coming from the glass. Since the "ray" coming out was new and unknown, it was called "X ray." Röntgen and others began using metals instead of glass in the path of the beam of electrons. For each different type of metal, a different X ray was produced. Some of the X rays had greater penetrating power than others.

Scientists then began to classify the X rays according to their penetrating power. The type that penetrated the best they called the "K series." The next best became the "L series" and then the "M series." The letters referred to electron shells in the atom. The experimenters discovered that the greater the atomic mass of the metal, the greater the penetrating power of the X ray that was produced. If this were true, then the nucleus had something to do with the way the electron beam was affected. Could something be discovered about the nucleus using this X-ray method?

Meanwhile, as Rutherford was becoming famous, he gathered into his laboratory a team of brilliant young men, a number of whom became famous in their own right. One was Henry Gwyn-Jeffrey Moseley (1887–1915). Moseley had graduated from Oxford in 1910 and had gone on to Manchester, England, where he studied under Rutherford. Moseley's father was a well-known zoologist. Moseley was a soldier in World War I and was killed by a Turkish sniper at the age of only twenty-seven.

Moseley worked in the laboratory with a friend, C. G. Darwin, a grandson of the great Charles Darwin. What Moseley did with the X ray was to discover exactly the kinds of rays produced by each element. He found that there was a direct relationship between the number of protons (plus charges) in the nucleus and the type of X ray produced. Moseley was then able to set all of the known elements in order according to the number of protons in the nucleus (atomic number). This fantastic piece of work by a young man was extremely useful in classification, as we shall see in a later chapter.

The picture of the "Rutherford atom" posed a problem. All physicists knew, according to the knowledge of that day, that when a charged particle like an electron traveled at great speed, it emitted energy. Now, if the electrons emitted energy, they would have to fall—fall right into home base, the nucleus. In fact, the physicists even described the way the electrons were supposed to fall: in a spiral motion. It was a hard problem to solve, and research on electronic arrangement in the atom was held up pending solution of the problem of the moving particle that wouldn't give up.

The Quantum Theory

Then, in 1913, the Danish physicist Niels Bohr (1855–1962) published a paper that said to all of the nuclear physicists, in effect: "Your physics tells you that the electron should radiate energy, and spiral into the nucleus. Since this is not the case, let us just say that it does not do so, and let us stop piddling around. Let us go ahead and make observations on the opposite assumption."

Bohr was the kind of scientist who made an assumption and then went about experimenting and gathering data to prove that assumption. On the other hand, some scientists first gather great amounts of data and then state the idea. In other words, Bohr talked before he proved.

Bohr's theory about the electrons in the atom was bold and new. Coupled with his name are those of many great mathematicians, all contributing to our present understanding of the arrangement of the electrons in the atom. We can mention the names of Louis de Broglie, Max Planck, Erwin Schrödinger, and Albert Einstein. There are many more.

The theory, started by Bohr, states that electrons carry tiny bundles of energy called "quanta," and the theory became known as the "Quantum Theory." One of the important ideas of the theory is that electrons travel in waves. Since this is so, the mathematical part of the theory is often called "wave mechanics."

The theory first states that the electrons are placed in the atom in definite energy levels, each at a different distance from the nucleus. The number of quanta possessed by the individual electron determines its level. The lowest energy level contains electrons with the least number of quanta. In order for these to fall into the nucleus, they would have to lose their quanta. Since these quanta are whole units, there is no such thing as "a tiny bit at a time." It would have to be "a whole unit at a time." The electron nearest the nucleus simply does not lose the whole quanta and therefore cannot fall in. We will expand further on the quantum theory in a later chapter.

III

Setting the Chemical House in Order

The several branches of science classify and set all of their material in order. If a science has no order, then it is no real science. The biologist classifies living organisms; he classifies the systems in which they function. He can inform you about digestion, circulation, reproduction. Everything has its proper place. The geologist has a scheme into which he fits his rocks; a test for hardness, a set of chemical formulas, everything just so.

What classification did the alchemist use? Classification? Are you joking? The alchemist liked to keep secrets; he probably didn't want anyone to know what a mess his "science" really was. There wasn't a decent book written about chemistry in those days. If an alchemist made any kind of discovery, he shielded it to gain favors and make money. He drew strange pictures to mystify; serpents with men's heads and strange necks and tails.

The ordinary people of the day were properly impressed. They could neither read the strange writings nor understand the stranger drawings. "What learning," they must have said, "what great intelligence these alchemists must have; they perform experiments with fire, and they write in strange and difficult tongues."

But really, what was there to classify in those days? The alchemist knew only a few elements, like gold, mercury, and sulfur. He did have one overriding factor. He had to support himself and his family, and for this he needed money. He had to buy equipment for his laboratory. His trade of alchemy did not earn money like the carpenter's trade, or the candlemaker's. Sometimes the alchemist would

FIG. 7. *The type of pictures that were drawn by alchemists*

obtain a patron, a rich man who would take a chance and hope that the alchemist would come across with some great discovery like changing base metals into gold, or inventing a medicine that could cure all diseases. For such hopes, the rich man would support the alchemist and his family.

The best alchemists would get the king himself as patron. This was good, and also not so good. If the alchemist didn't come through, he could lose his head.

The Alchemist's Nail

In order to make money, the alchemist often resorted to trickery. One trick that became quite famous was called the "nail." The unscrupulous alchemist would make a nail in his laboratory—a special one, half iron and half gold. He would then dirty it with mud.

Dressed in a long, sweeping gown, the wise man would go out, nail in pocket, into a crowded marketplace. "Ah!" he would suddenly exclaim, "I have found an old nail. I must try my recipe."

People stopped to look; a crowd quickly gathered. The alchemist cleaned that half of the nail that was iron, and the crowd saw the iron. Then:

"My recipe, my recipe. Ten years of hard work. My good recipe." He did not look right or left; he seemed not to know that people were everywhere. He groaned and moaned and cursed for some ten minutes. The crowd became larger.

Then, finally, he groaned his last groan, chanted his last chant, and took out from the depths of his cloak a vial of liquid. As he opened the vial, a foul sulfide smell was emitted that made everyone want to vomit. His half-closed eyes opening wider and wider, the alchemist finally thrust the nail into this foul liquid. He turned it, he twisted it, he thrust it in and out, and he talked in a strange tongue. Minutes passed; the great man finally ran the nail through an old rag. Seeming to be unaware of the crowd pushing around him, he held the nail at arm's length.

"Verily, verily, verily."

A stunned silence at first, then cries of:

"Sell me the recipe," "sell me the recipe," "I'll buy your recipe." The crowd had seen gold; with their own eyes they had seen the change from iron to gold.

We need say no more. The alchemist usually brought many vials hidden inside his great cape. And, yes, he often had to leave town very quickly.

There were honest alchemists, too. These even described some elements. They divided metals from nonmetals. Metals, they said, were shiny, could be hammered and bent, and conducted heat. The earliest metals known were silver, copper, tin, lead, and mercury. The earliest nonmetals were sulfur and carbon.

Triads of Elements

By the year 1865 many new elements had been discovered. Attempts were made to put some order into the awakening science. They were poor attempts, though, and failed to produce a good system. The best one at that time came from a German chemist, Johann W. Döbereiner (1780–1849), who placed some elements in groups because they behaved alike chemically and had similar physical properties.

The groups occurred in threes, so they were called "triads." One of the triads was chlorine, bromine, and iodine. The three gases combined alike with metals, and they were colored—very unusual for a gas. Chlorine was green; bromine, red; and iodine, purple. Another triad was calcium, strontium, and barium. All three combine in the same way with nonmetals, and all three give colors to a flame. Calcium is orange in a flame, strontium is red, and barium is green. Another triad was lithium, sodium, and potassium; still another was sulfur, selenium, and tellurium. It is interesting that our three famous magnetic elements, iron, cobalt, and nickel, formed a Döbereiner triad. Even today, the triads of Döbereiner are well known.

Döbereiner Triads

1	2	3	4	5
chlorine	calcium	lithium	sulfur	iron
bromine	strontium	sodium	selenium	nickel
iodine	barium	potassium	tellurium	cobalt

Döbereiner almost discovered a triad that included the element phosphorus. Phosphorus had been known since 1669, when a young

Setting the Chemical House in Order

chemist, Hennig Brandt, evaporated urine and discovered the element. What he did was to keep heating the dry matter that was left after gentle boiling. Most often, as he got down to the last bit, it burst into flames. That's the kind of an element phosphorus is—highly flammable. But, with patience, he finally succeeded and collected some pure phosphorus. He was amazed at how it glowed in the dark! He wrote about his discovery, and many alchemists were interested. One of these came to Brandt's house to buy some of the element. Brandt, needing money, was very willing to sell, but he did not have any phosphorus on hand. To make a supply would take too long; the customer was there and waiting. So Brandt came to an agreement with him to give him the raw material and the recipe. Brandt went into another room, and the customer, hearing sounds, guessed the nature of the raw material. Cursing under his breath, he left the house. He was last heard to mutter that alchemists were cheats.

About the year 1865, a Chemical Congress was assembled in Karlsruhe, Germany. The first and most vital issue was to try to establish a system for the study of chemistry. Classification of the elements was absolutely essential. It was this congress that strongly stimulated attempts to classify all known elements; to finally set the house of chemistry in order.

Efforts at Classification

One paper on classification had already been published by a Frenchman, Alexandre E. Beguyer de Chancourtois, in 1862. It was very badly received; in fact, it was practically discarded.

The publication of Beguyer described the elements as being arranged in a "Telluric Screw." The word "tellus" means "earth," and Beguyer meant "the screw of the elements of the earth." Elements that were known to be alike were placed in vertical lines. The entire scheme on paper was set on a cylinder, spiraling down. The triads of Döbereiner could be recognized. One trouble was that when the scheme was published the publisher forgot to include the important diagram. Without the diagram, the table was too difficult to understand.

It was sad that Beguyer had such a bad reception. The next scien-

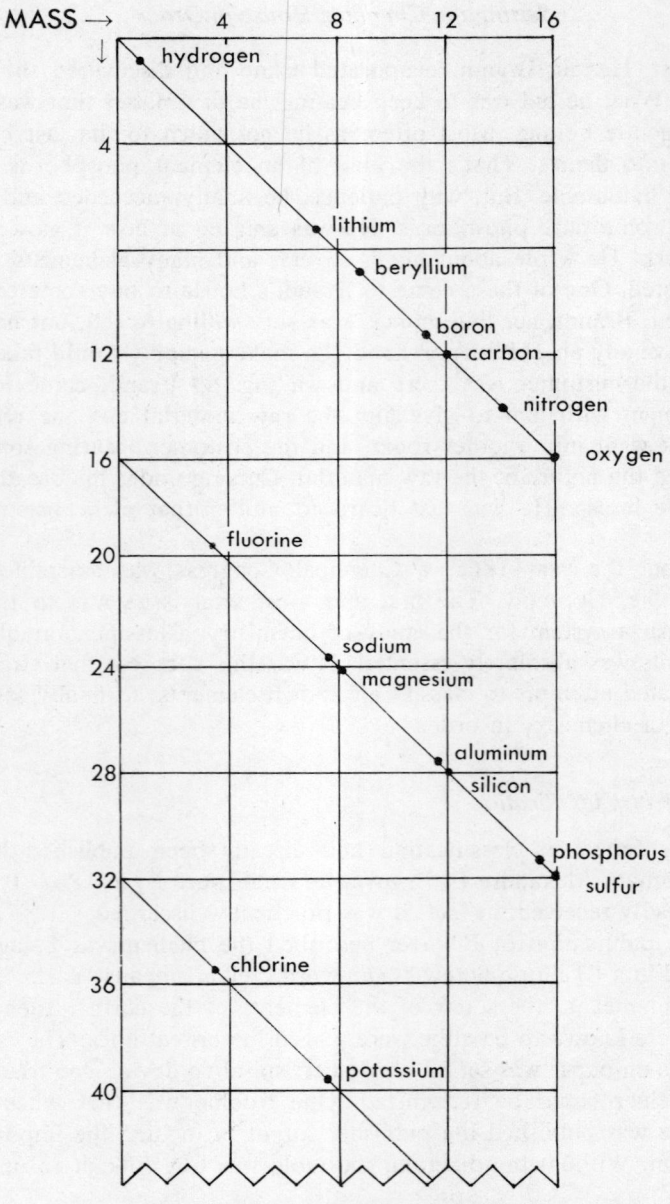

FIG. 8. *Part of Beguyer's Telluric Screw*

tist who tried had even worse luck. He was a young English chemist, John Alexander Newlands (1838–98). In 1864 he presented his classification paper before the British Chemical Society. Now, this society is a very conservative body, and was especially so in those days. Newlands had arranged all of the known elements in groups of seven. He noted that the eighth element was chemically similar to the first, and that this fact kept repeating itself. He called this type of elemental behavior the "Law of the Octaves." "Octaves" is a musical term, however, and as such it did not go over well in the field of science. In fact, it was ridiculed. His paper was not published.

Newlands kept fighting for his idea. He was greatly stimulated by the Chemical Congress of Karlsruhe. He thought that he surely had the answer, and he insisted on having his paper published. Finally, in 1887, he was allowed once more to present his paper before the society. This time it was accepted, and what was more, Newlands was awarded the important Humphry Davy medal.

In the portion of Newlands' table shown in Fig. 9, there are errors and omissions. First of all, he completely neglected column VIIIA.

H hydrogen	Li lithium	Be beryllium	B boron	C carbon	N nitrogen	O oxygen
F fluorine	Na sodium	Mg magnesium	Al aluminum	Si silicon	P phosphorus	S sulfur
Cl chlorine	K potassium	Ca calcium	Cr chromium	Ti titanium	Mn manganese	Fe iron

FIG. 9. *Part of Newlands' table*

Of course, the elements of that column had not been discovered at that time, but he could have indicated that something was missing, according to atomic masses. His table seemed to indicate that all elements in the world had already been discovered. He left no spaces at all for the possibility of new elements. Also, because he left no empty spaces, his known elements often fell into wrong places.

The Periodic Table

The next attempt at classification was made by two men at the same time, independently. Both men were absolutely brilliant. One was a German, Dr. Julius Lothar Meyer (1830–95), a gentleman and a scholar. Dr. Meyer was a chemist and a medical doctor who served as a physician during the Franco-Prussian war. As he worked with wounded soldiers, he tried to make their lot a little easier. He did a great deal of research, discovering, among other things, the effect of carbon monoxide on blood.

If Dr. Meyer had not been so busy as a physician, he might have devised a table of the elements sooner than he did. As it is, his table for classification of all known elements was an excellent one; but it was not quite ready when another one suddenly appeared. Dr. Meyer proved to be a excellent loser. He praised the other paper highly, saying that he could not have done so well. The truth is that his work was very similar to the one that was finally accepted by the world.

The other monumental work is the Periodic Table that is in use today (with some corrections). It was assembled by a Russian, Dmitri Ivanovitch Mendeleev (1834–1907). His father was a school principal who desired but one thing for his son, that he study the old languages and become a great scholar and teacher of language and literature. He recognized that Dmitri was exceptionally bright. Dmitri, however, was much more attracted to the sciences and mathematics than to languages. The father died suddenly while Dmitri was still a very young man. Dmitri took the exams for Moscow University and succeeded in passing. All students of the university are children of the government and are supported by the government. They are even provided with some "pocket money."

Moscow University is very proud of its student (and later teacher)

Setting the Chemical House in Order

Dmitri Ivanovitch Mendeleev. His tremendous task of assembling and classifying all of the elements proved to be a masterpiece. As far as is known, Mendeleev was not influenced by Newlands. The table that he finally published was wide and massive in its scope.

He set the elements in order of their atomic masses. In doing so, he had to find the best acceptable mass values for each element. He had to go over the work of other scientists with great care. Atomic masses were being determined by a ratio method. For example, a research chemist would note carefully the ratio with which elements combined to form a particular compound; then, by weighing the compound, he could determine the weight of each element. The fact is that several different weights could be reported for one element. Mendeleev had to determine which one was right.

Like Newlands, he found that after a certain number of elements, the chemical properties repeated themselves. Thus, he made "Periods," meaning that the properties were periodic. The table thus became known as the *Periodic Table*. There were horizontal rows called "periods" and vertical columns called "families."

Every so often, Mendeleev found that the next known element did not fit into the next available space. The chemical properties were different from those associated with the column. It was just wrong. Unlike Newland, he did not force the element to fit. Instead, he left the space empty, saying that an element belonged in the spot, but it had not yet been discovered. He even predicted chemical and physical properties for the unknowns. He was amazingly accurate in his predictions, and he spurred others to search for the missing elements. For example, the space under silicon could not be accounted for. Mendeleev actually named the still unknown element "ekasilicon" and predicted properties for it. Today we know the element as germanium, and its properties are very close to those predicted by the Russian scientist. The table shown is much like the original. You will find a great many similarities, but also a good number of differences compared to the table commonly used today.

The first column starts with the element lithium. Hydrogen, placed often in the first column, and sometimes in the seventh, does not fit anywhere. From lithium down, the elements are called "alkali metals." Having one rather "loose" electron in the outer layer, they are extremely active elements and very dangerous. They combine so

30 THE STUDENT CHEMIST EXPLORES ATOMS AND MOLECULES

R O W	I — R_2O	II — RO	III — R_2O_3	IV RH_4 RO_2	V RH_3 R_2O_5	VI RH_2 RO_3	VII RH R_2O_7	VIII — RO_4
1	H = 1							
2	Li = 7	Be = 9.4	B = 11	C = 12	N = 14	O = 16	F = 19	
3	Na = 23	Mg = 24	Al = 27.3	Si = 28	P = 31	S = 32	Cl = 35.5	
4	K = 39	Ca = 40	— = 44	Ti = 48	V = 51	Cr = 52	Mn = 55	Fe = 56 Co = 59 Ni = 59 Cu = 63
5	(Cu = 63)	Zn = 65	— = 68	— = 72	As = 75	Se = 78	Br = 80	
6	Rb = 85	Sr = 87	?Yt = 88	Zr = 90	Nb = 94	Mo = 96	— = 100	Ru = 104 Rh = 104 Pd = 106 Ag = 108
7	(Ag = 108)	Cd = 112	In = 113	Sn = 118	Sb = 122	Te = 125	I = 127	
8	Cs = 133	Ba = 137	?Di = 138	?Ce = 140				
9								
10			?Er = 178	?La = 180	Ta = 182	W = 184		Os = 195 Ir = 197 Pt = 198 Au = 199
11	(Au = 199)	Hg = 200	Ti = 204	Pb = 207	Bi = 208			
12				Th = 231		U = 240		

FIG. 10a. *The Periodic Table of Mendeleev*

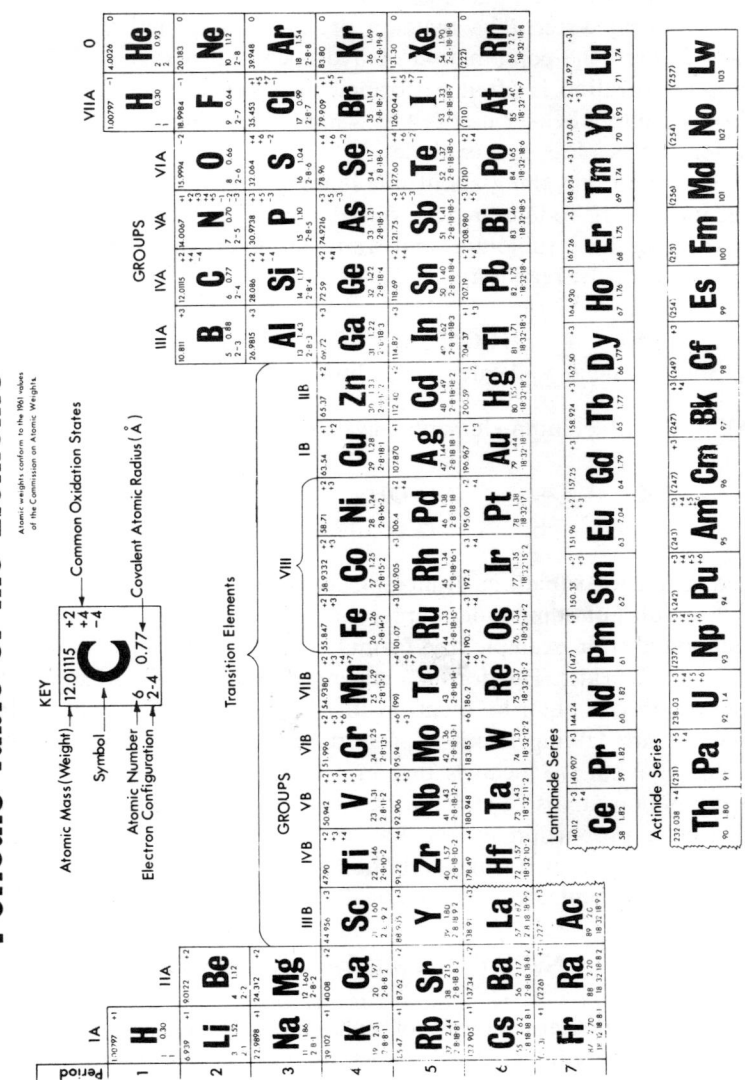

FIG. 10b. *The Periodic Table as used today in chemistry. Note that carbon is the element of reference; the atomic mass unit is now defined as 1/12 the mass of the carbon 12 isotope.*

readily with oxygen that they can knock a hydrogen atom out of the water molecule, and take its place. This reaction displaces hydrogen from water. The reaction produces heat, and the hydrogen catches fire (the combination of hydrogen with oxygen of the air). A demonstration is often performed in which a piece of sodium is dropped into a beaker half full of water. The sodium "dances" on the water, catches fire, and explodes.

The important thing about a column is that its elements act alike chemically. If sodium reacts with water, then so do lithium, potassium, cesium, rubidium, and francium. One can write a chemical equation for the reaction:

$$2\,Na + 2\,H\,OH \rightleftharpoons 2\,Na\,OH + H_2$$
Sodium metal　　Water　　Sodium hydroxide　　Hydrogen gas

All of the IA column elements have the same equation:

$$2\,K + 2\,H\,OH \rightleftharpoons 2\,K\,OH + H_2$$
$$2\,Cs + 2\,H\,OH \rightleftharpoons 2\,Cs\,OH + H_2$$

and so on for rubidium, francium, and lithium.

Formulas of compounds represent the ratio of the elements in the compound. For example, the formula of common table salt is NaCl, sodium chloride. The formula means that one sodium atom combines with one chlorine atom, in that proportion. We shall see later that we are talking about a proportion, and not a unique particle "NaCl." What we mean is that in a crystal of sodium chloride, for every billion atoms of sodium, there are a billion atoms of chlorine. If such a proportion is true for sodium, then it must be true for lithium, potassium, cesium, rubidium, and francium. Every time you learn one formula, you automatically learn six:

NaCl, KCl, LiCl, CsCl, RbCl, FrCl.

Here is another example of column unity. The second column starts with beryllium. The formula for beryllium chloride is $BeCl_2$. Without any question, we know the formulas for the other members of the family:

$BeCl_2$, $MgCl_2$, $CaCl_2$, $BaCl_2$, $SrCl_2$, $RaCl_2$.

Chemistry made easy! Thank you, Dr. Mendeleev.

In the next chapter, we shall discuss the "bonding" of the elements to form compounds. Meanwhile, it is important to recall the Bohr idea of the atom. We ended Chapter II with the Bohr concept. According to this concept, the electrons are arranged in energy levels around the nucleus. To understand the idea of a "family" of elements is to appreciate the importance of the outermost electrons in an atom. These are the "valence" electrons, and they determine the chemistry of the element. They determine the bonding power of each individual atom. Thus, in column IA, all of the elements in the family have one single electron in the valence layer. In column IIA, the valence layer has two single electrons. The whole family, all with the two unpaired electrons, act the same way chemically. The story continues with columns IIIA, IVA, VA to VIIIA.

Because of our awareness that the valence electrons constitute the most important factor in the chemistry of the elements, we realize that the order imposed on the elements in Mendeleev's table should be the order of protons. It is the number of protons that dictates the number of electrons. Both numbers are exactly the same in the neutral atom. What then did Mendeleev actually do? He used the weights of the elements, which included the neutrons. The fact that he came out mostly right is because of the normal increase in neutrons as the element gets bigger. Once in a while, he had to go wrong. The neutrons do not have to add exactly in order.

Here is an example of error because of using atomic masses (weights). The weight of tellurium is 127.50; the weight of iodine is a little less, 126.905; thus, iodine would go first on a weight table. This would cause tellurium to fall into column VIIA. Column VIIA is a very important column, a family of nonmetals including fluorine, chlorine, bromine, (iodine), and astatine. In no way does tellurium, a metalloid type of substance, fit in. Tellurium does fit under the VIA column, after oxygen, sulfur, and selenium. And, using atomic numbers (number of protons), we find that tellurium has an atomic number of 52, and iodine has an atomic number of 53. These are correct.

Mendeleev was clever enough to know that iodine belonged in

column VIIA. In studying the various reported weights obtained in experiments, he often got a number of different weights for one element. What he did for tellurium and iodine was to use weights that would make them fit into their proper places. The weight he used for tellurium was not the correct one, 127.50, but a reported weight of 125. This lower weight made it fall into the right place. Mendeleev was correct in his final placement of the two elements, but the way he achieved it was capricious or unscientific. In fact, research workers have a name for this type of thing: "fudging."

In Chapter II we referred to the work of Moseley with the X ray. You remember that when a stream of electrons penetrate an element, X rays are emitted. Moseley determined that for each atomic structure, the X rays emitted were different. His theory was that, when the beam penetrated, it actually penetrated deep into the atom. The innermost electrons of the atom were forced out. Other electrons dropped down to take the places vacated. In fact, it was this dropping down that caused the X rays. Whenever electrons go from a higher level of energy to a lower level of energy, they lose a definite amount of energy. This lost energy is the X ray. Moseley discovered that there was a direct relationship between the X rays given out and the pulling power of the nucleus. This pulling power of the nucleus is really due to the number of protons in the nucleus. By many, many bombardments with electron beams, he was able to set the Periodic Table in the order of atomic numbers instead of atomic weights. Although there were not many changes from the original table, the basic idea was that it was the most correct way of arranging the elements. After all, chemistry today is the result of accumulated knowledge. It is the result of many experiments by many scientists, never the result of just one man's experiments or ideas. It is team work that counts in the long run.

IV

The Marriage of Elements

"Marriages are made in heaven." This must certainly apply to chemical marriages, since atoms fit so perfectly when joined together. In fact, atoms of different elements form a union (compound) that has new properties all its own. It is often quite difficult to break the union apart.

It is essential to know the difference between a compound and a mixture. A mixture is not a marriage. Instead, it is a fickle, temporary relationship. Each partner retains his own independence and his own properties. The partners can often be separated with ease. For example, suppose we mix table salt (mostly sodium chloride) and iron filings. The two can be separated by putting the mixture into a glass of water. The table salt will dissolve, while the iron filings will fall to the bottom. We can then filter out the filings (and even clean them on the filter with fresh water) and, in order to recover the table salt, evaporate all of the water.

Compounds

What then is a compound? The table salt itself, sodium chloride, is a union of two elements, sodium and chlorine. Sodium is a very active metal and a powerful poison. The nonmetal, chlorine, is a gas; it was used in World War I as a poison gas. Both are bad news. Yet, put them together, and lo, we have a substance that is vital in our diet and makes food tastier. When on safari in Africa, I saw

36 THE STUDENT CHEMIST EXPLORES ATOMS AND MOLECULES

elephants travel miles just to lick some salt. Many animals seek sodium chloride. It is a good example of the great changes in properties that occur when elements combine to form compounds.

The fact that atoms can combine should not be new to us. After all, very few elements are found free, as individual atoms, in nature. Take the oxygen of the air. Two atoms of oxygen join to form the molecule O_2. (A discussion of the meaning of "molecule" will follow soon.) Metals are rarely found as free metals; rather they are found combined with nonmetals. The most common nonmetals in these combinations are oxygen, sulfur, and chlorine. Carbon exists mainly in combinations with hydrogen, with oxygen, and with nitrogen. Carbon exists in a great number of combinations. It actually forms the basis of life itself. Clearly then, the combined state is most natural.

In the 18th century, many chemists were trying to break compounds apart to discover the elements themselves. In 1756 a student at Edinburgh University, Joseph Black (1728–99), preparing a paper for his doctoral degree, broke up the compound calcium carbonate. Calcium carbonate is found as limestone, marble, and pearls. Its formula is $CaCO_3$. This compound can be broken down into two parts, calcium oxide (CaO) and carbon dioxide (CO_2). Carbon dioxide is a gas, and gases were hardly known in those days. All gases were called "airs." When Black heated calcium carbonate, this new "air" was emitted. A small piece of marble, just an inch cube, gave enough gas to "fill a vessel holding six wine gallons." There was great amazement at this information.

Examination of the Periodic Table can give clues as to the way in which atoms combine. The designations IA through VIIIA at the top of each column tell us the number of electrons in the outermost layer of the elements in the column. For example, in the IA column, lithium, sodium, potassium, etc., all have one electron in the outer layer. Thus, all elements in column IA act alike, chemically. They combine with a nonmetal in the same proportion. They combine with oxygen in the ratio of 2:1. The formula for lithium oxide would then be Li_2O; for sodium oxide it would be Na_2O; potassium oxide, K_2O; and so on for the rest of the elements in the column. For the purpose of explaining combinations of elements better, we move next to column VIIIA.

The "Noble" Elements

In column VIIIA we find the elements helium, neon, argon, krypton, xenon, and radon. These six elements have one unique property. They remain alone. They are not attracted to and will have nothing to do with other elements. It is true that some pretty tricks have been performed to show that it is possible for column VIIIA elements to form compounds. Krypton, xenon, and radon were forced to combine with fluorine, and xenon also combined with oxygen. These are truly exceptions, however, and we must think of this particular column as being "noble" and aloof.

Helium, the first element in the column, has two protons in its nucleus. With the two protons, there must be two electrons to make a neutral atom. These two electrons travel in space around the nucleus, a space called "orbital." The two make a pair, and as a pair have no tendency to combine with electrons from other atoms. The secret of the lonely helium atom is this pair of electrons. It is the unpaired electrons that are wanted and sought after.

The other elements of column VIIIA have eight electrons in their outer layer. These eight make a perfect four pairs. Thus, the elements of column "perfect" are unique. They remain as they are. All other elements in all other columns have at least one electron that is unpaired.

Now, let us reexamine the compound sodium chloride. The sodium atom has eleven protons in the nucleus, and therefore eleven electrons dashing around outside. The first layer of electrons has one orbital and a pair of electrons, just like helium. The next layer holds four orbitals, and therefore eight electrons. Up to this point there is a resemblance to VIIIA elements; first helium and then neon. Up to the neon structure, there are ten electrons. Now, the one remaining electron places itself in a new shell and is an unpaired, ready-to-react electron.

Since the real chemistry lies with the outermost, single electron, we can show the element sodium in a simple way, as:

$$\mathbf{Na^{\cdot}}$$

Fig. 11a.

38   THE STUDENT CHEMIST EXPLORES ATOMS AND MOLECULES

The element chlorine has seventeen protons in the nucleus. It therefore has seventeen electrons moving in orbitals around the center of the atom. The closest layer has the helium two, the next layer has the neon eight. There are seven more to go. The seven have four orbitals; the electrons must arrange into three pairs and one unpaired electron. The one unpaired electron is a determining factor in the chemistry of chlorine.

$$:\overset{..}{\underset{..}{Cl}}\cdot$$

FIG. 11b.

Now, putting the sodium atom and the chlorine atom together, using the dot system:

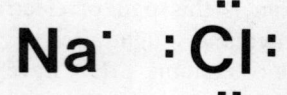

$$Na\cdot \quad :\overset{..}{\underset{..}{Cl}}:$$

FIG. 11c.

When the two are put together, the two single electrons join and become a pair. What, then, happens to the newly formed pair? How do they arrange themselves in this brand-new situation?

In order to answer the question where the united pair finally make their home, we must return to the beauty of perfect column VIIIA. The perfection of this column and its great stability is imitated by all other atoms whenever possible. In the case of chlorine, as it combines with sodium, if the pair of bonding electrons were held close, it could resemble argon, an element in the royal column. A stable, happy condition.

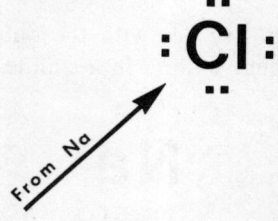

$$:\overset{..}{\underset{..}{Cl}}:$$

↑
From Na

FIG. 11d.

The Marriage of Elements

Under these conditions, what happens to sodium? What happens to it when its one lonely, newly paired electron departs to keep house at chlorine? The sodium atom now remains with ten electrons. Ten electrons is the exact arrangement of yet another noble gas, neon. Thus, by a giving and a taking, a certain stability is achieved.

At the same time, the atoms of chlorine and sodium are not the same atoms they were before. They are altered. Chlorine, previously with seventeen electrons, now has eighteen. This puts a stress on the seventeen-plus nucleus; it cannot hold eighteen-minus adequately. The entire chlorine atom then gets fatter, the electrons moving farther away from the nucleus. More than just physical, the atom now is unbalanced with charge. It has a minus in excess. A particle that has a charge is no longer called an atom. It is called an "ion," and the chlorine ion is written:

$$Cl^-$$

Fig. 11e.

The sodium atom is in exactly the opposite situation. The eleven-plus nucleus no longer has eleven electrons to hold; instead it has ten. One has deserted to chlorine. The nucleus pulls the weakened ten toward itself, and the entire atom becomes much smaller. But, again, we cannot call the new arrangement of the sodium an atom any longer. It has lost its neutrality, and it has one plus in excess. It is therefore an ion, and is written:

$$Na^+$$

The situation is stable. The sodium ion exerts a pulling force toward its lost electron. This pulling force is part of a force field that is commonly called "electrostatic attraction." Electrostatic attraction holds the ions together in an ionic compound.

With ionic bonding, a network or "lattice" is formed by the ions. In such a lattice, the plus ions are surrounded by minus ions, and the minus ions are surrounded by plus ions. Wherever possible, like charges will stay as far from each other as they can. You have seen very beautiful crystals; ruby, sapphire, amethyst, opal, and others. In these can be found lattice arrangements of ions. Try growing copper sulfate.

40   THE STUDENT CHEMIST EXPLORES ATOMS AND MOLECULES

Sodium Chloride

● Chlorine (Ions)
○ Sodium (Ions)

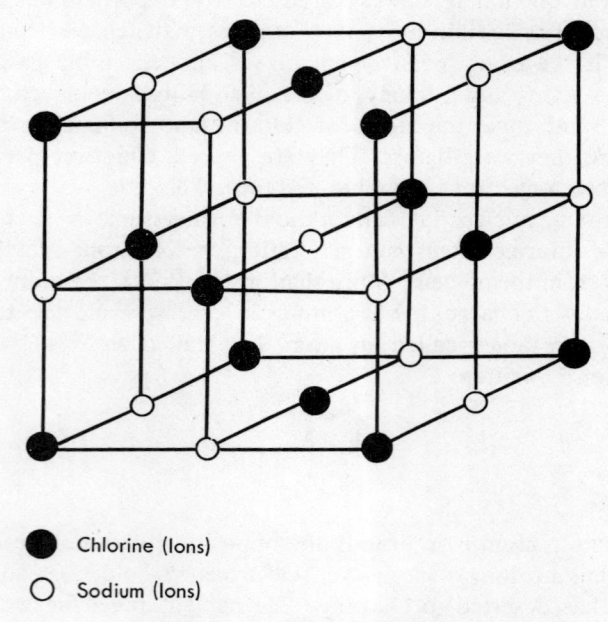

FIG. 12a.   *Electrostatic attraction*

The sodium chloride crystal is itself a cube shape.

FIG. 12b.   *Sodium chloride crystal*

It is easy to point out that there is no single particle that one can call "sodium chloride." In other words, what we have under ionic conditions is ions surrounding ions. One cannot distinguish units of NaCl; we are unable to take the compound apart and have pieces of NaCl. That is why it is not a molecule.

To see how molecules are formed, we have to illustrate yet another type of bonding. In this case we combine a nonmetal with another

The Marriage of Elements 41

nonmetal. Recall that ionic bonding was a metal-nonmetal affair. The easiest to illustrate would be hydrogen. One hydrogen atom can join with another hydrogen atom:

$$H\cdot \ \cdot H$$

Fig. 13a.

The tendency to form such a pair arrangement, much like helium of VIIIA, is responsible for the bonding. After the union, each hydrogen atom shares the pair. There is no reason why one should have the pair more than the other. With such perfect sharing, a neat little bundle is formed.

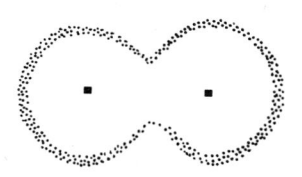

Fig. 13b.

There is no crystal lattice; there is just a unit in space called hydrogen gas. This is a molecule. The same type of union would occur with two atoms of oxygen. The molecule N_2 is even more interesting. The nitrogen atom has three unpaired electrons.

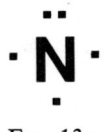

$$\cdot \overset{\cdot\cdot}{\underset{\cdot}{N}} \cdot$$

Fig. 13c.

When two nitrogen atoms combine, their electrons are bonded as shown.

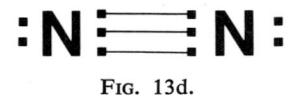

$$:N\equiv N:$$

Fig. 13d.

Triple-bonded nitrogen is difficult to break apart. It has more than triple the strength of a single bond. This stability is fortunate for the living organisms that breathe. These organisms would be endangered if the nitrogen molecule opened into atoms. Nitrogen atoms have unpaired electrons, and unpaired electrons react too well for comfort.

Nonmetals resemble the VIIIA octet much more than do metals. The important valence layer (the outside layer) has four to seven electrons and is ready to be filled up to eight. The easiest, of course, is the case of VIIA; it needs only one. Also, the smaller these nonmetals are, the better. The nucleus is not so shielded by electrons in the small element, and, after all, it is the plus nucleus that has to do the attracting for other electrons. The larger an element is (going down a column, for example), the less able it is to "fill up." Since this ability to fill a valence layer is a distinguishing characteristic of nonmetals, large elements (in the same column as nonmetals) often lose the nonmetal character.

The carbon column starts with a nonmetal and ends with tin and lead. The nitrogen column starts with a nonmetal and ends with metallic bismuth. Once more, the oxygen column ends with the metal that the Curies discovered, polonium. The actual number of nonmetals is quite small:

hydrogen	sulfur	chlorine	bromine
carbon	nitrogen	fluorine	iodine
oxygen	phosphorus		

The type of bonding between nonmetal and nonmetal is one in which the newly formed electron couples are "shared." Since such sharing involves the valence layers getting together, the name given to such sharing is "covalent." Covalent bonding makes molecules. Molecules, as we mentioned, are distinct units. An example is water. The structure of water is known to be:

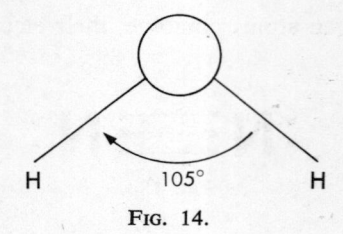

FIG. 14.

The Marriage of Elements

Electronegatives

In order to discuss the water molecule in more detail, we have to look at a set of numbers. The American chemist Linus Pauling (1901–   ) invented this set of numbers to show the force of attraction that each element has for electrons; not its own electrons, but outsiders. At the top of the list is fluorine; no element exceeds fluorine in pulling electrons. Next comes oxygen and the nonmetals. The metals come last, column IA being the lowest of all. With the help of these numbers, we can easily tell whether a bond is ionic or covalent. More than that, however, we can also see small differences in how much the bond leans toward being one or the other.

Let's go back to sodium chloride with the numbers of Pauling. The numbers are called "electronegativities." The electronegativity of sodium is 0.93. The electronegativity of chlorine is 3.16. We know, by the chemical and physical behavior of the salt, that it is ionic. To find the type of bond we are dealing with, we subtract the smaller number from the larger one as follows:

$$\begin{aligned} \text{chlorine} &= 3.16 \\ \text{sodium} &= \underline{0.93} \\ \text{bond value} &= 2.23 \end{aligned}$$

Little by little, through knowledge about thousands of compounds, chemists have decided that a difference of 1.7 or higher means that an ionic bond is formed. With such a difference, one of the combining atoms is able to grab the bonding electrons almost for itself.

What would the bonds in water be like, using this method of electronegativities? The electronegativity of oxygen is 3.44, and that of hydrogen is 2.20. Therefore:

$$\begin{aligned} \text{oxygen} &= 3.44 \\ \text{hydrogen} &= \underline{2.20} \\ \text{bond value} &= 1.24 \end{aligned}$$

The oxygen-hydrogen bond is a covalent bond. There are covalent bonds that tell one story, however, and covalent bonds that tell another. Observe, once again, our picture of a water molecule:

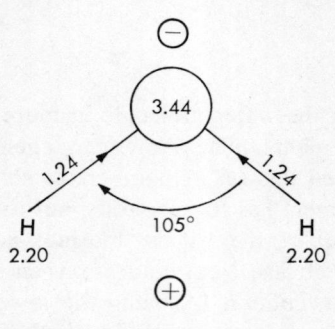

44 THE STUDENT CHEMIST EXPLORES ATOMS AND MOLECULES

```
         ⊖
       ┌───┐
       │3.44│
       └───┘
      1.24   1.24
   H    105°    H
  2.20          2.20
         ⊕
```

FIG. 15.

This time we have introduced the numerical value for the bond. If the value were less than 1.24, the pairing electrons between oxygen and hydrogen would not go so close to oxygen. Sharing is still taking place, but it is uneven sharing, favoring the oxygen. Because of this "favoring," one end of the molecule is richer in electrons and the other end is relatively poor in electrons. This sort of situation is called "polar." The water molecule is said to be "polar covalent."

The fact that water is polar covalent leads to another type of bond. This type of bond is associated so much with molecules of living organisms that we like to think of it as a "living" bond. In the case of water, this bond holds several water molecules together at room temperatures. The fact that several molecules of H_2O are held together makes water a liquid at these temperatures. The molecule itself is small and light, and as individual particles would no doubt be a gas. The polar nature of the tiny particle makes it act like a little magnet (Fig. 16).

Now, in your biology class, you have learned about proteins. You may also have touched on DNA molecules and RNA molecules. All of these have the type of bond described. It is called a "hydrogen bond." The attachment is always between a hydrogen and either an oxygen or a nitrogen. It can be between a hydrogen and a fluorine, but that is much less common. In the case of water, the bond is intermolecular—between molecules. In the case of the tremendous structures of the living tissues and living functions, the bonding is most often inside of the molecule itself, shaping it (Fig. 17).

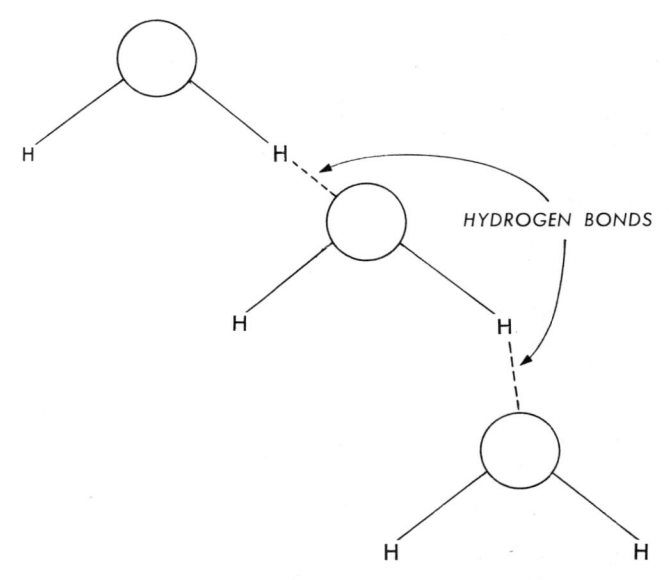

FIG. 16. *Polar nature of water molecule*

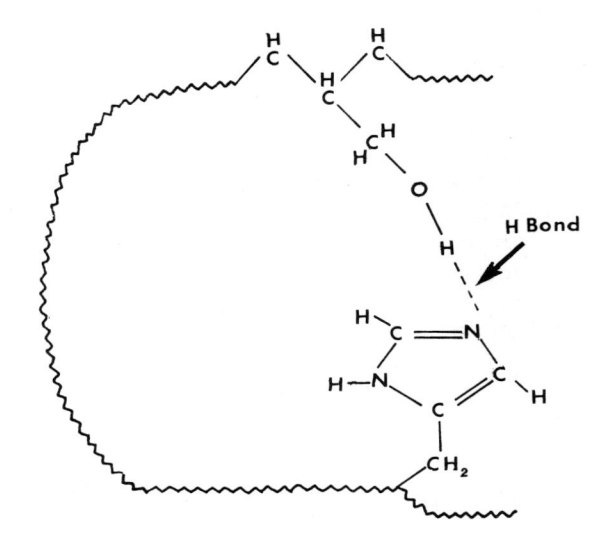

FIG. 17. *Hydrogen bond*

46 THE STUDENT CHEMIST EXPLORES ATOMS AND MOLECULES

Since there is such a thing as a polar bond, what is a "nonpolar" bond? It is a bond that has an equality of charge; it is also a molecule that is nonpolar because it has symmetry. H_2 is nonpolar. O_2 and N_2 are nonpolar. A molecule like carbon dioxide is nonpolar, although it is made of two polar bonds:

$$O = C = O$$

The C—O bond is polar, but when two are put together and the angle between is a straight line of 180°, a nonpolar effect is obtained. Both ends are the same. To be polar, one end must be more electron-rich than the other. Another example of a nonpolar molecule is methane, CH_4.

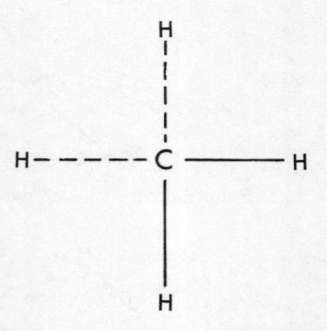

```
          H
          |
          |
          |
          |
H - - - - C ———— H
          |
          |
          |
          H
```

FIG. 18. *Methane molecule*

The individual carbon-hydrogen bonds are each polar-covalent. Carbon and hydrogen are, after all, different elements, and each has a different electronegative value. Carbon pulls electrons a little better than hydrogen does. But, when the polar-covalent bonds are arranged in a symmetrical pattern, the entire molecule is nonpolar.

Unless they have a high molecular weight, nonpolar molecules are often gases, or liquids that become gases very easily. Nonpolar molecules may even be solids if they become very large:

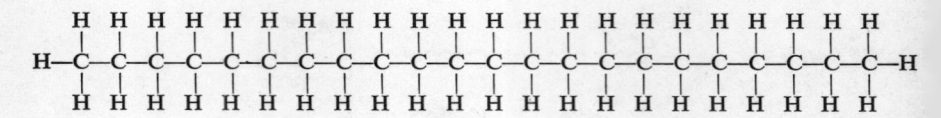

```
     H H H H H H H H H H H H H H H H H H H H
     | | | | | | | | | | | | | | | | | | | |
H—C—C—C—C—C—C—C—C—C—C—C—C—C—C—C—C—C—C—C—C—H
     | | | | | | | | | | | | | | | | | | | |
     H H H H H H H H H H H H H H H H H H H H
```

The Marriage of Elements

Nonpolar molecules like carbon dioxide, if cooled to a low temperature, become liquid or solid. In the case of carbon dioxide, it becomes solid at about $-77°$ C. As a solid it is commonly known as "dry ice."

We have seen that polar molecules attract each other and thus form liquids or even solids. But how is it possible for the same thing to happen to nonpolar molecules? The question brings us to another type of intermolecular bond.

The carbon dioxide molecule has twenty electrons; six for each carbon, and eight for the oxygen. We cannot take for granted that at all times, at any particular fraction of a second, all twenty electrons are arranged around the atom in an exact symmetrical fashion. There must be times when, for reasons of chance, one end will have more electrons than the other. When this takes place, there is really a little bit of polarity. Furthermore, neighbor molecules are affected by this slight polarity. If one end becomes plus, electrons from a neighbor are attracted, and that end of the neighbor becomes minus:

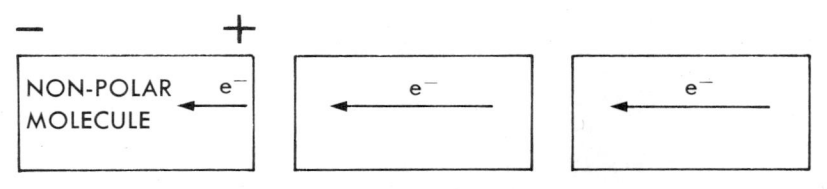

FIG. 19. *Induction effect*

Cooling brings the neighbors close together. As a result, the "induction" effect is felt more and more. Trillions and trillions of particles begin to stick together. Such pulling together or bonding is due to van der Waal's forces.

Another type of bond of which all chemists are aware is called a "coordinate covalent" bond. The main thing about this bond is that it really is a covalent bond; the difference lies in the way it is formed. It does not form by unpaired electron meeting unpaired electron. Instead, it uses a pair of electrons already there. Take the case of the water molecule:

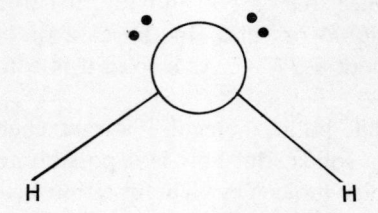

48   THE STUDENT CHEMIST EXPLORES ATOMS AND MOLECULES

FIG. 20.   *Coordinate covalent bond*

We have accented the pairs of electrons sticking out into space above the oxygen atom. If some particle happened along that had a completely empty orbital, it could very nicely sit down on one electron pair. A covalent bond would be formed. It is simply a case of "have pair, will share" and "have none, want some." We can use a hydrogen atom that has lost its one electron as the "have none, want some" partner in the bonding:

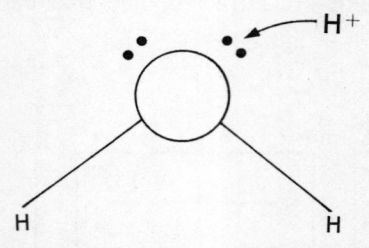

FIG. 21.   *Covalent bond with hydrogen ion*

The bond, after it forms, is no different from any other covalent bond. The story is told only in the way it actually happens. There are many coordinate covalent bonds, and they form because a pair of electrons are exposed and a substance having an empty orbital comes along. The reactant with the pair of electrons is called a "base," and the other reactant with the empty orbital is called an "acid."

The last type of bond in our discussion is the "metal" bond. Metals always have one or more empty orbitals. This means that the valence electrons can jump. They can jump all over the place. Atoms of metals are arranged next to each other. The jumping electrons visit other atoms. In fact, they can no longer be identified with the original atom they came from. They go very fast, through the entire metal system of atoms. The atoms, minus their moving electrons, become charged plus. The metal bond now becomes plain. The sea of moving electrons are, at the same time, held by attraction to the plus atoms. Thus, they are like billions of cords held tightly to the package. The metal bond is a powerful one.

The moving electrons give the metal some of its characteristics. The electrons move in waves that capture light waves. All of the light waves (or almost all) are then radiated back. The metal assumes a high luster. If one end of the metal is heated, the electrons move even faster. The entire metal object is affected, and the whole thing becomes "hot."

In the case of electricity, an additional load of electrons are fed into the metal. The additionals are past the neutral point. No more electrons are needed by the metal. Either the new electrons move through empty orbitals, or they push other electrons down the metal by repulsion. In any case, the metal bond causes the metal to allow a flow of electricity.

V

How Strong Is a Chemical Wedding?

It happens mostly because of "singles parties" (unpaired electrons). Meetings are cleverly arranged. The parties in question are pushed close together, and they join. Some of these marriages are very strong; some are much weaker. The weak ones change partners as soon as they can. The relative strength of these joinings is usually well known by the chemist. He has lists that tell him the quantity of heat he must use to break a particular bond. Very often, the chemist can break an ionic bond by putting the substance into water. The ionic substance and water react, and usually enough heat is generated by such a reaction to split the bond. This type of reaction is based on "solubility," and a knowledge of solubility is vital to the work of the chemist. Before we discuss solubility, let's look at the relative strengths of bonds when they are subjected to direct heat.

Heats of Formation

Some of the cleverest people in the world are detectives. It is remarkable how a good detective ferrets out the reasons behind a crime. The same is true of the clever chemist. If a bond is to be broken, he wants to know how it was formed and all about its possible associations. When atoms join, heat is emitted. This is called an "exothermic" reaction. In order to break the bond, it becomes necessary to put back at least the same amount of heat. Clearly, the more heat there is emitted in the formation, the stronger is the bond. A table called "Heats of Formation" is, therefore, a

How Strong Is a Chemical Wedding?

table that tells about the strength of bonds. What sort of experiments were done to arrive at "Heats of Formation"? One type of experiment is the "bomb."

A solid steel "bomb" that can be closed very tightly is used. Into this are placed two elements. Both are carefully measured, by weight or volume. Let's say the two elements are hydrogen and chlorine, both gases. The bomb has an electrical connection, and a spark can be produced inside. After the bomb is filled, it is put into a tank of water. However, the tank is not an ordinary one. Like a refrigerator, it is well insulated. It has a very delicate thermometer that can measure temperatures with great accuracy. When everything is set, the chemist pushes the switch. After the explosion inside the bomb, the change in temperature is observed. When everything has been taken into consideration, the heat of formation of hydrogen chloride is carefully calculated. A very good indication of the strength of the new bond has thus been determined.

Heats of formation are common in many things going on around us. The rusting of iron, for example. I look sadly at my poor iron fence, which is showing rust in many places. The iron is slowly changing its strong metallic bond for another type of bond. The secret of the lion-strength in steel is the metallic bond, and once it loses that, it becomes as weak as a kitten. Meanwhile, what happens to the heat of formation? Looking at the table, I find that a lot of heat should come out when rust is formed. Yet, when I touch the fence and the rust spots, I feel no extra heat.

There are two reasons why the heat of formation cannot be felt. One reason is that only a relatively small amount of the metal is actually exposed to air oxygen. The reaction of a metal and oxygen would be very fast if powdered metal were used. Finely powdered metal in the air could cause an explosion if a match were lit. The iron fence has only a limited surface exposed to the air; therefore, the oxidation reaction is slow.

The other reason why the heat of formation cannot be felt is that our bodies are not sensitive enough. We cannot feel very small changes in temperature. These small changes caused by the reaction going on in the fence add up through the years. They equal, all added together, the amount of heat that would come out if the reaction were very fast.

Strengths of Bonds

The iron-oxygen bond is an ionic bond, a metal to nonmetal bond. There is a very good experiment that illustrates the difference in strength between the ionic and covalent bonds. The experiment uses a compound like potassium chlorate ($KClO_3$). You may have used this chemical in your chemistry lab.

$KClO_3$ has both types of bonds. It has an ionic bond where the metal potassium is connected to ClO_3. It has covalent bonds between chlorine and oxygen. If the $KClO_3$ were to be heated, which bond would break first?

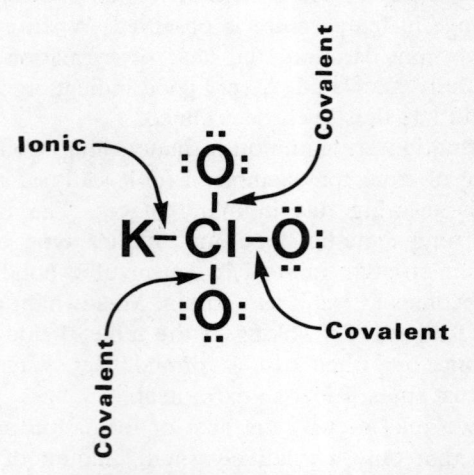

FIG. 22. *Structure of potassium chlorate*

The potassium chlorate salt is put into a test tube, and the test tube is clamped in a holder on a ring stand and heated until the salt melts. The heat is kept going. If wooden splints are put into the test tubes, they burst into white-hot flames that shoot into the air. (Hazard—this experiment may be dangerous; be very careful to stand at a good distance from the flaming material and point the test tube away from you.) There is no doubt that pure oxygen is being released from the potassium chlorate. The spectacle is great when the lights are put out. Soon, the splints do not catch fire any

How Strong Is a Chemical Wedding?

more; the oxygen is gone. The powder remaining in the test tube is tested and found to be potassium chloride (KCl). Of course, potassium chloride, metal to nonmetal, has an ionic bond. The covalent bonds have been devastated, wrecked, shattered by the heat. The ionic bond is left intact. In fact, if the KCl were to be made hotter and hotter, you would be amazed at how strong its electrostatic bonds really are!

What about the metallic bond? Our skyscrapers rest on pillars of steel. "I" beams of steel support the millions of tons of concrete that make some streets in the cities look like canyons. What is the secret of the metallic bond that makes for such tremendous rigidity? The piece of steel that you hold in your hand—what is it like inside? What could you, yourself, do to answer such a question. I have done an experiment with my students in class. Taking a small metal box, I have placed some small objects in it and sealed the cover. Then, giving it to a student, I have said,

"Tell me what's in the box, without opening it."

"But, how can I?" At first, the student is completely disturbed by the problem. You can easily guess what happens next. The box is shaken, listened to, vibrated, dropped, studied. Some good answers emerge.

"There are more than two pieces inside. They are metallic. Some of them are round, because they roll."

The piece of steel can also be investigated, much like the closed box. Observation of its luster and its ability to conduct electricity and heat are evidences of roving electrons. The actual binding strength of these electrons can be measured by machines commonly used in engineering. The tensile-strength machine pulls the piece of steel until it breaks. This certainly measures the stength of metallic bonds.

An interesting machine "chops" a piece off the metal. The steel is held tightly in the machine, with a few inches sticking up into the air. Then comes a great swinging hammer that lops off a few inches. A large numbered face shows how much the hammer was forced to slow down when it hit the metal. The amount of slow-down indicates the strength of the metal (the shear strength) against being broken apart. Again, this is an indication of bond strength.

We come back again to the contest between the ionic and the covalent bonds. We have indicated that the winner in strength is the

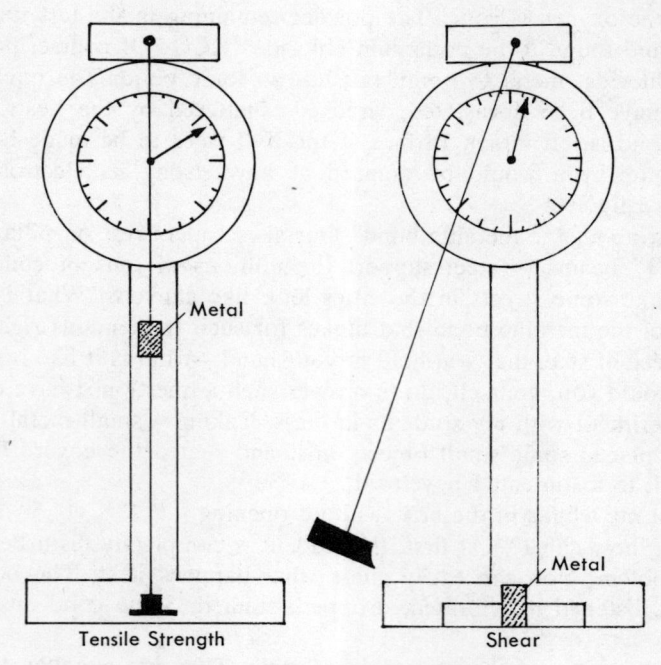

FIG. 23. *Testing strength of bond*

ionic bond. Suppose we add ionic sodium chloride to water. It dissolves. It breaks into the tiniest pieces, much like gas molecules in the air. We cannot see the one, nor can we see the other. This means, of course, that the sodium ions and the chloride ions have come apart. What! Can this be our strong ionic bonding?

Now, put a covalently bonded substance in water. Oil is a good example. Placed in water, the oil floats on top. The covalent bonds do not separate. Does this mean that covalent bonds are stronger than ionic bonds? Not at all!

Ionic compounds enter into a reaction with water. We have seen in Chapter IV that the water molecule is polar: it has a minus end and a plus end. When ionic chemicals enter the domain of water, great attractive forces are created. The metal ion is plus, and the minus end of water is attracted to it. As many water molecules as can fit bombard the metal ion. This gives off heat. Meanwhile, the

How Strong Is a Chemical Wedding?

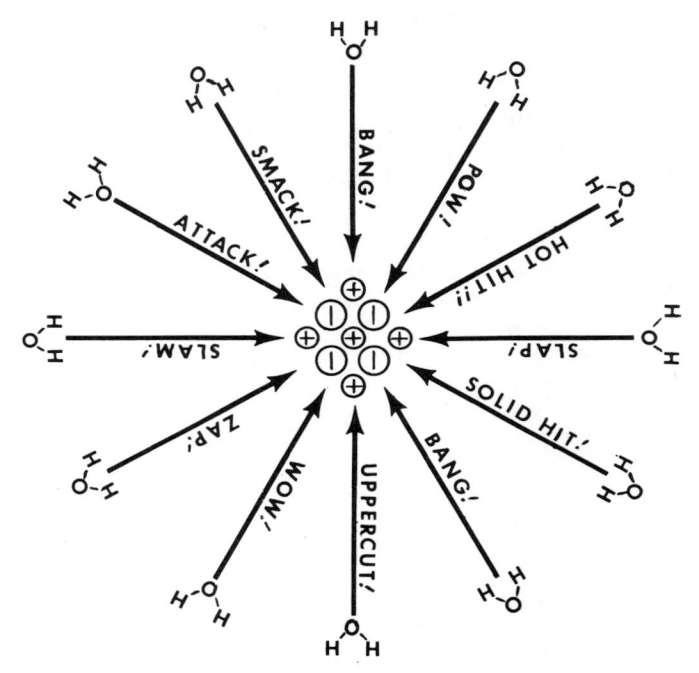

FIG. 24. *NaCl under attack in water*

plus end of the water molecule is attracted to the nonmetal ion. Heat, again, is given off. In order for the ionic bond to break, enough heat must come out of these reactions. The heat must also break some of the hydrogen bonds between water molecules. Small units of water must fit properly around each ion. These units must take each ion for a swim.

The heat produced by the water-hitting-ion reaction
EQUALS
the heat to break the bond *PLUS* heat to break hydrogen bonds.

With such information it becomes easy to see why some ionic compounds are soluble in water and some are not. Sometimes, because of the way the ions are packed, the ions are harder for water to get at. Sometimes, too, because of stronger types of ionic bonding, the hydration energy just isn't enough to break the bonds. For example, the element barium has two valence electrons, and therefore a two plus charge. It can form strong attachments to double-charged minus ions (like sulfate, SO_4^{--}). These attractions are much more difficult to break than attractions between single charges. Thus, many barium compounds are only slightly soluble in water.

If oil or benzene or kerosene were put into water, they would not dissolve at all. These molecules, being covalent, have very little plus-minus character. The polar water molecules do not attack. The covalent bond will not break.

VI

The Remarkable Aspects of Carbon and Silicon

What a fantastic element carbon is! It is soot, it is a gleaming diamond, it is a graphite rod. It is the basic element of almost all "living" compounds, the thousands of compounds that make up living organisms. It is the basic element of medicines, of dyes, of plastics, of gasoline, and of oil. In the Periodic Table, carbon is placed in the very center of period 2, horizontally across the page. The placement itself calls for unique properties.

Starting with column IA, lithium has only one electron in the valence layer, and loses it very easily. Beryllium, next in line, has two electrons in the valence layer. It takes more energy to take these off than in the case of lithium. As we go along the row, there is greater and greater reluctance to give up electrons. Nitrogen, in column VA, begins to take electrons rather than give. Oxygen almost always takes, whereas fluorine always takes electrons, and never gives any.

	I	II	III	IV	V	VI	VII	VIII
R O W 2	Li Lithium	Be Berylium	B Boron	C Carbon	N Nitrogen	O Oxygen	F Fluorine	Ne Neon

FIG. 25. *Row 2 of Periodic Table*

Between this giving and taking lies column IVA, and especially the element carbon. Because of its position, carbon's compounds are covalent; neither giving nor taking, but "sharing." Compared to the other atoms on the Periodic Table, the second-row atoms are small. The real difference between carbon and its sister element silicon, directly below, is that carbon is smaller. Carbon, because of its smallness, can form chains of carbon atoms:

$$C-C-C-C-C-C-C-C$$

Silicon, although it has the same number of valence electrons as carbon, cannot form stable covalent bonds with other silicon atoms because of its size.

The Carbon Compounds

Let's examine carbon more closely. The atom has four unpaired electrons in the valence layer. It can, then, make four connections with other atoms. The other atoms must also have unpaired electrons. The unpaired electrons around carbon travel in orbitals. It is thought that one orbital has a different shape than do the other three. The different one has a travel shape called "s," and the three that have the same shape in travel are said to describe a "p" orbital.

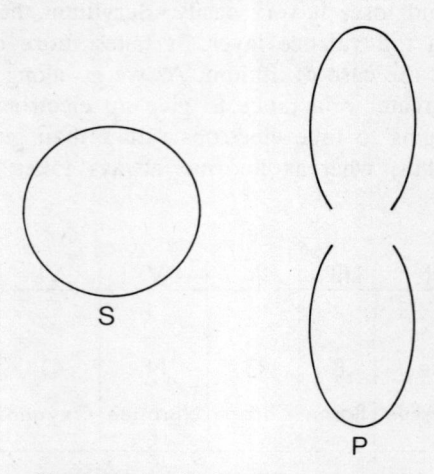

FIG. 26a.   *Orbitals of carbon atom*

The Remarkable Aspects of Carbon and Silicon

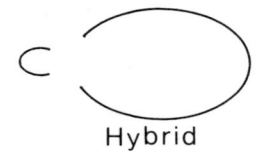

Hybrid

FIG. 26b. *Hybrid orbital of carbon atom*

Covalent bonds, the kind that carbon forms, force a geometry on the structure of the molecule. We find that, although the carbon atom seems to start out with uneven shapes in its valence layer, the geometry of its final compounds does not have uneven shapes around it. It must be that nature blends together the round and the dumbbell shapes and makes one final shape for all.

Further, we find that a molecule like methane, CH_4, a combination of four hydrogen atoms with one carbon atom, has the bonds coming away from carbon in a regular tetrahedron. The angles that form in such a four-sided figure are angles of 109°.

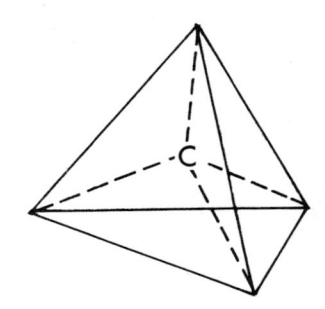

carbon tetrahedron

FIG. 27a. *Carbon tetrahedron*

In order to see more plainly the geometry of carbon compounds it is important to know the 109° angle. For example, in the case of "cyclic" structures, the angles that are formed may impose a strain on the natural tetrahedral angle.

There is not much strain in the five-membered ring. The natural angle for such a ring is 108° between carbons. In the six-membered ring, however, the angle forced on each carbon-carbon bond would be about 120°. This is too much of a strain, and the structure must

60 THE STUDENT CHEMIST EXPLORES ATOMS AND MOLECULES

cyclopropane

cyclohexane

cyclopentane

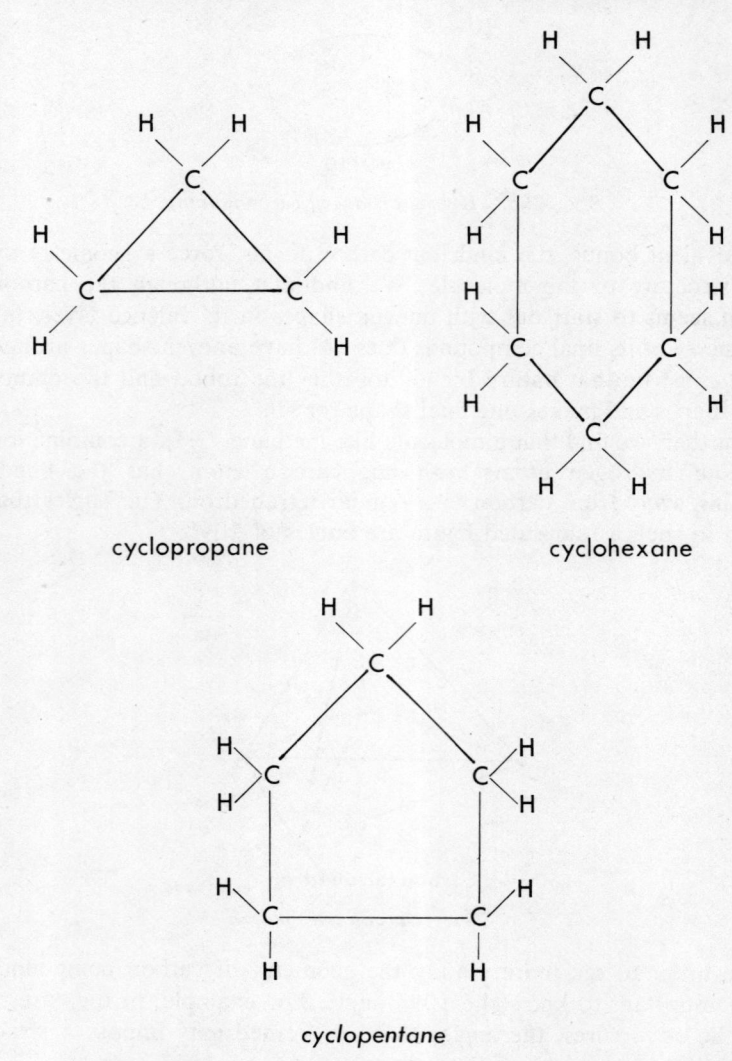

FIG. 27b. *Left to right: cyclopropane, cyclohexane, cyclopentane*

bend to relieve the strain. Two forms of the six-membered ring are possible:

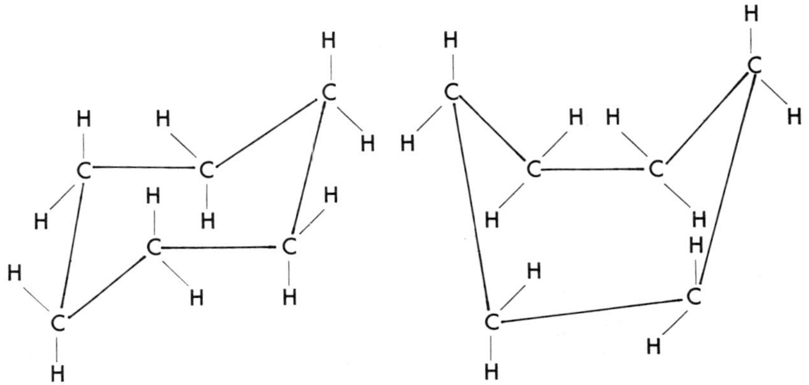

chair form   boat form

FIG. 27c. *Cyclohexane: left, chair form; right, boat form*

The ring is called "cyclohexane." It is a very interesting structure because so many compounds have it. Many chemists refer to cyclohexane as a "puckered" compound. Ball-and-stick models are very useful in demonstrating the cyclohexane structure. With the help of such a model, one can easily show the change from one shape into the other. One end is simply rotated into the other shape. The chair form is strongly favored because it permits the hydrogen atoms to go as far apart as possible. A compound that is very common to us is the simple sugar, glucose. The structure of glucose resembles that of cyclohexane:

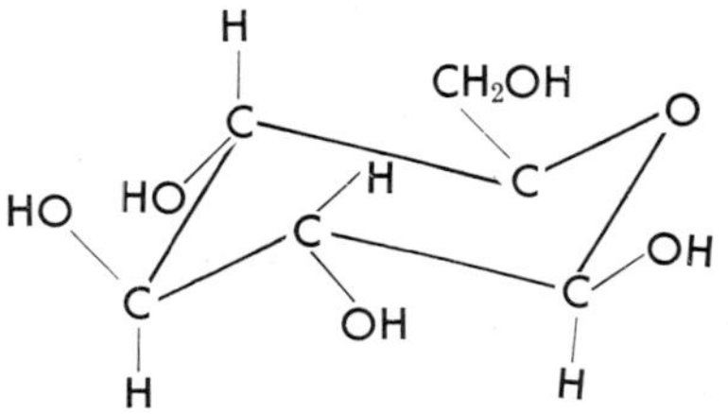

glucose

FIG. 28. *Structure of glucose*

62   THE STUDENT CHEMIST EXPLORES ATOMS AND MOLECULES

Many elements bond with carbon. One of the most common is oxygen. An example of carbon-oxygen bonding is the compound ethanol:

$$\begin{array}{c} \text{H} \quad \text{H} \\ | \quad\;\; | \\ \text{H—C—C—O—H} \\ | \quad\;\; | \\ \text{H} \quad \text{H} \end{array}$$

Chemists prefer to write the "structural" formula of ethanol as shown above. It is poor chemistry to write the formula of this alcohol as C_2H_6O. It is true that the compound contains two carbons, six hydrogens, and one oxygen. The fault in not using the structural formula lies in the fact that another and different compound, methyl ether, (C_2H_6O), can be written:

$$\begin{array}{c} \text{H} \quad\;\; \text{H} \\ | \quad\;\;\;\; | \\ \text{H—C—O—C—H} \\ | \quad\;\;\;\; | \\ \text{H} \quad\;\; \text{H} \end{array}$$

In carbon chemistry, the actual structure of a compound showing the way atoms are attached to each other is most important. Two compounds like ethanol and methyl ether are called structural isomers of each other. Larger compounds can have many isomers, since there are many ways to arrange covalent bonding. Carbon compounds have more versatility than others.

Double Bonds

Carbon compounds have another kind of versatility: they can have different "types" of bonds. Up to now, we have discussed four separate bonds made by carbon. It is possible, however, for carbon to make two bonds to another carbon.

This kind of bond is called a "double" bond. The molecular structure shown is that of ethylene. In chemistry the ending of a compound name is important. In this case, the "-ene" ending means

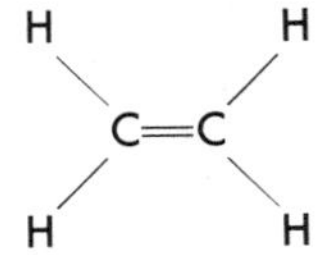

$$\text{H}\diagdown_{\text{H}}\diagup^{\text{H}}\text{C}=\text{C}\diagup^{\text{H}}\diagdown_{\text{H}}$$

ethene (ethylene)

FIG. 29a. *Ethene (ethylene)*

a double bond. Ethylene is different from its single-bond companions. The carbon angles change from 109° to 120°. This is not a good arrangement, nor a stable one. The entire compound attempts to return to the tetrahedral 109° angle. It thus becomes very easy for another substance to add on to ethylene:

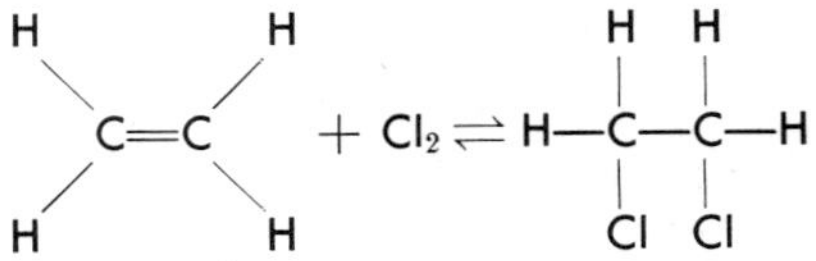

$$\text{H}_2\text{C}=\text{CH}_2 + \text{Cl}_2 \rightleftharpoons \text{H--CHCl--CHCl--H}$$

ethene   1,2,dichloroethane

FIG. 29b. *Ethene transformed into 1,2,dichloroethane*

Because double-bond compounds are always ready to add and become "saturated," such compounds are called "unsaturated" compounds.

Double bonds can also be found between carbon and oxygen, or carbon and nitrogen. Acetone is a carbon-to-oxygen double-bond substance, and is a very common solvent.

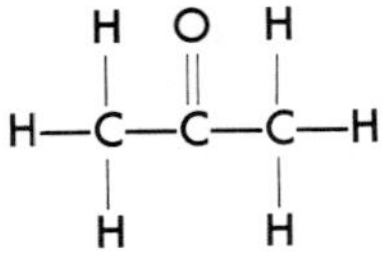

$$\text{H}_3\text{C--CO--CH}_3$$

acetone

FIG. 29c. *Acetone*

64 THE STUDENT CHEMIST EXPLORES ATOMS AND MOLECULES

Acetone is used as a solvent to dissolve many kinds of large organic compounds. In the geometry of double-bond structures, the double bond flattens the compound. Because of this flattening effect, another possibility of isomers exists. Let's take an example using two chlorine atoms on ethylene:

$$\begin{array}{cc} H & H \\ | & | \\ C = C \\ | & | \\ Cl & Cl \end{array} \qquad \begin{array}{cc} H & Cl \\ | & | \\ C = C \\ | & | \\ Cl & H \end{array}$$

"cis" form   "trans" form
cis dichloroethene   trans dichloroethene

FIG. 29d. *Left, cis dichloroethene; right, trans dichloroethene*

In the formula at left, the two chlorine atoms are on one side of the double bond. In the formula at right, the two chlorine atoms are on opposite sides. Again and again we see the tremendous variety in carbon compounds!

Now we come to the triple bond between carbon atoms. Three of the unpaired electrons from one carbon pair up with a corresponding three from another carbon atom:

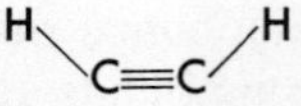

$$H-C\equiv C-H$$

ethyne

FIG. 29e. *Ethyne*

Such a molecule is called "ethyne." Again the ending of the word indicates that the nature of the molecule is to have a triple bond. It is "-ane" for a single bond, "-ene" for a double bond, and "-yne" for a triple bond.

Ethyne (commonly called acetylene) is even more unsaturated than ethylene. It is commonly used in the oxyacetylene torch, and was once used as a fuel in miners' lamps. The very unsaturated nature of the compound makes it combine quickly with other atoms. It can be very dangerous.

The Remarkable Aspects of Carbon and Silicon

Finally, we come to a type of carbon molecule that has alternating double bonds and single bonds:

$$C-C=C-C=C-C=C$$

These alternating double bonds are very interesting chemically and are the basis for making many compounds, among them hormones of the human body, and rubber. One compound that has often been illustrated as a conjugate system (another name for alternating double bonds) is benzene.

benzene form C_6H_6

FIG. 30a. *Structure of benzene*

Benzene is a very special compound. It is a ring structure with three double bonds. It is a very stable, unreactive substance. You will recall that the double bond is actually very fast to react, able to add many kinds of atoms to itself. This conjugate system does not seem to be the same. In benzene there is no double-bond addition.

Examination of the second bond in each case shows that it is due to one single, unpaired electron from each carbon. In fact, in the benzene ring, six electrons have been responsible for the double bonds. These six electrons actually form an equal sharing partnership throughout the entire benzene structure.

benzene—sextet form

FIG. 30b. *Sextet form of benzene*

The "sextet" of electrons stabilizes the entire ring structure. There are no real double bonds, hence there are no double-bond addition reactions. The number of benzene ring structures in organic chemistry is so vast that chemists have given those compounds a special name, "aromatic compounds." All other compounds are called "aliphatic compounds."

Silicon reacts the same way as carbon. The valence layer has the same number of electrons. In silicon, however, the valence electrons are farther away from the nucleus. Silicon is a larger atom than carbon. The valence electrons are also farther apart from each other. Because of these conditions, the silicon-to-silicon bond is not nearly as strong as the carbon-to-carbon bond. Long chains of silicon will not form.

The most stable bond that is common with silicon is the silicon-oxygen bond. Oxygen can easily replace most other elements that are attached to silicon:

$$Si\,H_4 + O_2 \rightleftharpoons SiO_2 + 2H_2O$$

The compound SiO_2 (silicon dioxide) actually forms a lattice structure much like the brilliant diamond. We know silicon dioxide as quartz, and also as sand. If it is melted and cast, it becomes glass.

Although Si-Si bonds are weak, the Si-O bond is very strong. Long chains can be made with the two elements. These long, complicated compounds are called siloxanes, or more commonly, "silicones." They are truly giant molecules, and we shall have more to say about them later.

VII

How Small Molecules Grow into Giants

Not long ago a Russian scientist reported that a substance called "polywater" had been made in his laboratory. What did he mean by the name? "Poly" means "many"; thus "polywater" would be a substance composed of many water molecules. In fact, it would be a giant molecule. Many very large molecules had been prepared before, but polywater was unique. Certain very exact conditions were necessary to make it. Ordinary water had to be heated in closed, thin glass tubes for a long time. The product was a thick gelatinous fluid.

Scientists were unsure just how the water molecules attached themselves to one another. Water is a polar substance, and has an area of negative charge on the oxygen, and a positive charge on the hydrogens. It is through these charges that water holds together in the liquid and solid phases. The bond is called a hydrogen bond and, in itself, is not very strong. We have mentioned this type of bond earlier. It is what makes water unusual—a molecular substance that sticks together.

The structure of polywater still baffled the world. Chemists felt that there was no adequate explanation of how the entire giant molecule stuck together. Weak hydrogen bonds just could not do it. After all, liquid water itself was already hydrogen-bonded. There was every reason to believe that water could not stick together tightly as a large number of H_2O molecules. Yet, there was this thick liquid that looked like Jello.

Finally, the saga of "polywater" ended very sadly. It was proven to be nothing more than a solution of silicon dioxide in water. The thickening of the water occurred by its dissolving some of the glass, largely silicon dioxide, when heating was done at high temperatures. Not mentioned in early papers was the need to use freshly made glass tubes in the preparation. The explanation of polywater was now quite clear. A layer of silicon dioxide was removed from the glass. Silicon dioxide has a great tendency to form a gel with water under high temperatures. After one layer of glass was removed, no more would come out. Therefore, no more polywater would form. This marvelous substance passed into history as a famous chemical hoax.

Giant Molecules

Fortunately, the history of large molecules does not have many such incidents as polywater. It was in 1920 that Hermann Staudinger published a paper that began speculation about the making of giant molecules. He stated that life occurred only because small molecules joined together to form big ones. The bonds in these giants were the same covalent bonds that held the small ones together. These covalent bonds joined all of the small ones together in intermolecular fashion as well as they had within the molecule itself.

Back in 1828 a young German chemist had accidentally disproven the important "vital force" theory, which said that substances derived from the life process had a particular "something" that made it impossible to duplicate in the laboratory. Friedrich Wöhler needed a chemical called ammonium cyanate. Ammonium cyanate is a substance that most certainly has nothing to do with the life process. In making it, the atoms rearranged themselves into a well-known molecular material called "urea." There is no doubt that urea is part of the living kingdom of molecules.

Staudinger had drawn attention to the importance of giant molecules in nature. Since the time of Wöhler's experiment, "molecules in nature" also meant molecules in the laboratory. The substances that Staudinger pointed to were very unusual and contained many thousands of atoms for each unit molecule. They consisted of a series of very small units, all joined together. In other words, these giants of nature had regular structures. Each single, small unit was

How Small Molecules Grow into Giants

called a "monomer," and was joined to other units by similar links. Thus many monomers appeared in a regular fashion to form a giant molecule called a "polymer." Furthermore, polymers could be broken down into their individual units by careful chemical treatment. The most common natural giant molecules are the carbohydrates, the proteins, and the nucleic acids. The monomer in many carbohydrates is the compound glucose.

Glucose units can be connected to form a polyglucose. Two examples of polyglucose are starch and cellulose (cotton and wood). These are things we see around us every day.

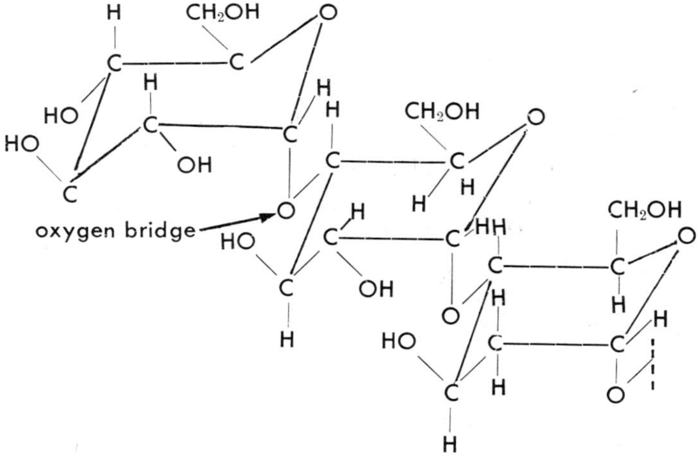

FIG. 31a. *Polyglucose*

Proteins, polymers of living organisms, are large and have many forms. The monomers of proteins are the amino acids. And, finally, nucleic acids, the substances of heredity, are made up of a unit that consists of two compounds and phosphate.

The realization that large molecules were really made of repeated small units was a stimulus for new research. Through research, the plastics industry in the United States was developed. One of the important ideas in the laboratory was that the double bond between carbon atoms might be able to reorganize itself. In the reorganization, it could form new bonds with other molecules.

```
    H       H   H       H   H       H
     \     /     \     /     \     /
..... C = C   +   C = C   +   C = C  .....  ⇌
     /     \     /     \     /     \
    H       H   H       H   H       H
```

ethylene monomers

```
      H   H   H   H   H   H
      |   |   |   |   |   |
..... C — C — C — C — C — C .....
      |   |   |   |   |   |
      H   H   H   H   H   H
```

FIG. 31b. Polyethylene

The product of such addition reactions between ethylene monomers is called "polyethylene," a plastic we all recognize. It is the main constituent of garden hoses and plastic containers. The long chain polymer is not really as straight as in the example above. It is usually produced as a crude plastic, and in the process of formation some small branches get stuck on the "main line."

Now, regular chains like the "linear" polyethylenes can nestle together and form a sort of crystal:

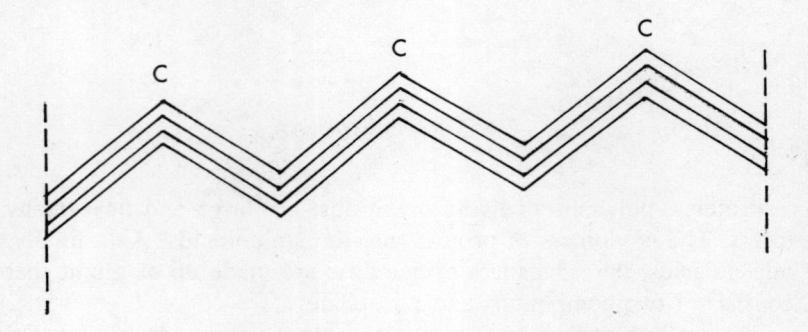

FIG. 31c. Nestling

The branched material cannot do this, and does not stick together nearly as well. The result of not sticking together is a low-melting, gummy substance.

Suppose you want certain properties in a plastic; you want to manufacture squeeze bottles. You need a flexible material. What could you, as a fine polymer chemist, do about making such a plastic? Well, you start thinking about the monomer. Ethylene, if allowed to polymerize so that it could nestle, would have a too rigid structure; certainly not squeezable. Then, in order to avoid some of the nestling, it might be possible to upset the pattern some. Substitute some other group instead of one of the hydrogens. This would not allow as much nestling and would make the material more pliable. If a methyl group (CH₃) were used, the plastic would not be called polyethylene any longer, but instead would be "polypropylene."

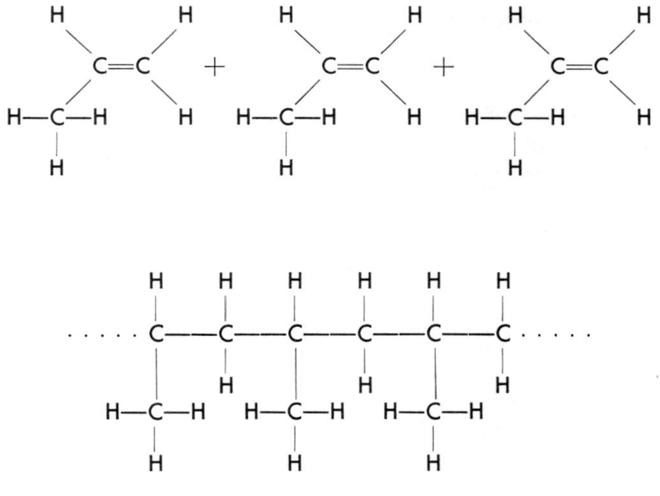

$$\begin{array}{c}H\\ \diagdown\\ C=C\\ \diagup\quad\diagdown\\ H-C-H\quad H\\ |\\ H\end{array} + \begin{array}{c}H\\ \diagdown\\ C=C\\ \diagup\quad\diagdown\\ H\quad H-C-H\\ |\\ H\end{array} + \begin{array}{c}H\\ \diagdown\\ C=C\\ \diagup\quad\diagdown\\ H-C-H\quad H\\ |\\ H\end{array}$$

FIG. 31d. *Polypropylene*

It is interesting to note that if we use a benzene ring on the original ethylene, we get so much interference in the nestling that the "polystyrene" is actually a "liquid." In fact, it is a glue, rubber cement. Further polymerization makes it a solid.

A very similar type of giant molecule is one that makes up rubber. Many novels and nonfiction accounts have been written about the

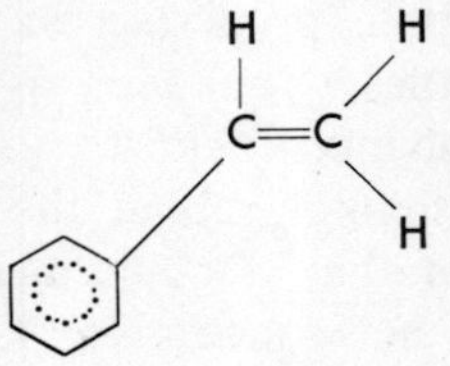

FIG. 31e. *Styrene*

rubber plantation, the rubber industry, and how rubber came to be used in the first place. The first real use of rubber was in the Amazon jungles of Peru. The people there "tapped" the hevea tree and obtained a thick liquid that they called "caoutchouc." They spread this liquid on cloth and on shoes. The liquid forms a polymer and hardens into a soft, flexible material. The Amazonians had thus succeeded in waterproofing some of their clothes. When the stuff was first taken to England, the chemist Joseph Priestley got hold of some. He said, "It is excellently adapted to the purpose of wiping from paper the marks of a black lead pencil." The material he referred to had come from the West Indies, and it was thereafter called "India rubber."

Vulcanization of Rubber

Until 1839 rubber was something of a curiosity, used in making toy balls, erasers, and not much else. In 1839 Charles Goodyear had an accident. He had been experimenting for about ten years, trying to do something important with rubber. He accidentally spilled a mixture of latex (the natural rubber) and sulfur on a hot stove. Up to this point, he had been unsuccessful and was ready to admit failure. This spilled mixture of latex and sulfur probably stank. Goodyear finally scraped the mess off the stove, noticing that it had become firm. He didn't throw it away, but waited for it to cool. Then he examined it. It was much harder than before, and it still had an elastic feel. It looked like what he had been trying to make! He tested it at different temperatures. He tried its resistance to acids and bases. It was an excellent material. Charles Goodyear had discovered "vulcanization." It was the beginning of automobile tires, bowling balls, and battery cases.

But what did he actually do to the raw rubber? The basic compound for rubber is isoprene.

$$\begin{array}{c}H\\ \diagdown\\ \diagup\\ H\end{array} C{=}C \begin{array}{c}CH_3\\ |\\ |\\ H\end{array}{-}{-}C{=}C \begin{array}{c}H\\ |\\ |\\ \diagup\\ \diagdown\\ H\end{array} \; + \; \begin{array}{c}H\\ \diagdown\\ \diagup\\ H\end{array} C{=}C \begin{array}{c}H\\ |\\ |\\ H\end{array}{-}{-}C{=}C \begin{array}{c}CH_3\\ |\\ |\\ \diagup\\ \diagdown\\ H\end{array} \begin{array}{c}H\\ \\ \\ H\end{array}$$

isoprene

↓

rubber segment with CH₃ groups and double bonds between C2–C3

+
sulfur

↓

Two rubber segments cross-linked by sulfur bridges (S–S and single S)

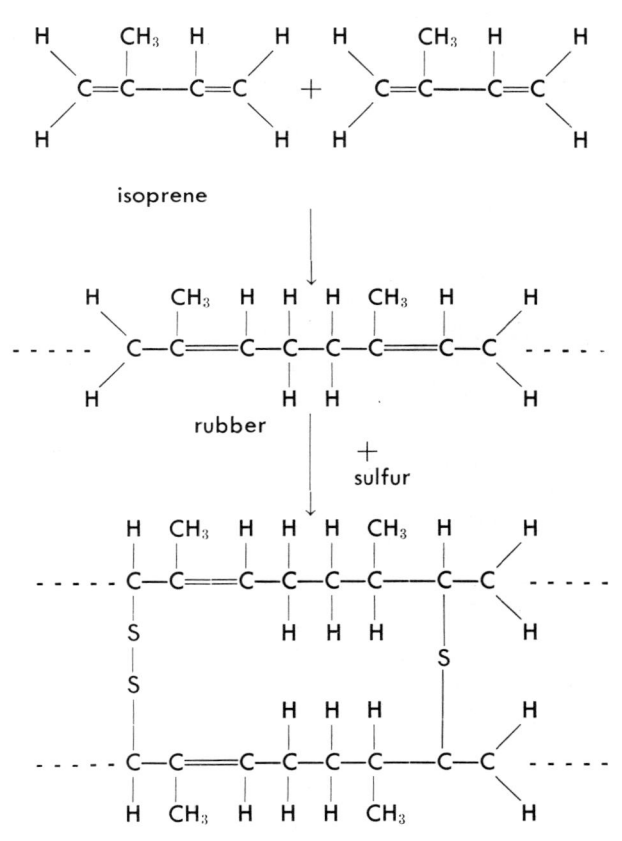

FIG. 32. *Two types of sulfur cross-links*

When isoprenes combine, like the joining of the ethylenes, the result is a new double bond between carbon two and carbon three. When sulfur is added, two long segments can be linked by sulfur bridges. In order to make the right kind of rubber for automobile tires, the chemist must not use too much sulfur. If all the double bonds are cross-linked by sulfur, the material obtained is much too hard. In that case, it could be used to make bowling balls and battery cases.

The story of Charles Goodyear does not have a happy ending. His idea and work were stolen, and he fought very hard to regain his rights. He died a poor man, leaving his family with many debts.

Synthetic Rubber

Rubber now became more and more important to the world. The United States bought raw rubber from Southeast Asia and vulcanized it in factories at home. Then suddenly, on that infamous day in December, 1941, Pearl Harbor was bombed. The ships from the rubber plantations of Southeast Asia could not come in. The war was on the verge of being lost. So much of our war effort depended on rubber! The U.S. government sent a desperate plea to the organic chemists of the country: Make rubber!

The organic chemists responded.

It is interesting to turn to the enemy at this point. What was he doing about rubber? The Americans could successfully block the Nazis' natural source of rubber, and they did. But Germany had already learned to make synthetic rubber and did not care about the blockade. The Germans had anticipated trouble before we did. They knew about isoprene in latex, and they made a number of similar molecules. For example, they made a compound called butadiene:

$$\begin{array}{cccc} H & H & H & H \\ C\!\!=\!\!C\!\!-\!\!C\!\!=\!\!C \\ H & & & H \end{array}$$

Polymerization reactions are exactly the same with butadiene, but different properties result because of the absence of the methyl (CH_3) group.

These and other synthetic rubbers were so successful that they are in wide use today. Much of the rubber produced in wartime was actually very poor in quality, since it was still in the experimental stage. Today, synthetic rubbers are better than natural rubber in many respects. For example, natural rubber deteriorates in the presence of oil. There are, however, oil-resistant synthetic rubbers.

We have thus far discussed the simple addition reaction of small molecules. These all involve the joining together to form a long chain polymer. There is another method by which giant molecules form. This method is used by living organisms to make starches, proteins, and other body substances. It is also used by the chemist to make thousands of kinds of plastics. Nylon and Bakelite are made by this method. In these reactions, water molecules are removed; therefore they are called "condensation" reactions.

In the case of the starches, when condensation occurs, an oxygen bridge is formed. We have seen an illustration of polyglucose in a previous chapter. In polyglucose there are a number of —OH groups on the ring structures. It thus becomes possible for side oxygen bridges to form.

$$\text{_____OH} \quad \text{HO_____}$$

If a water molecule leaves, an oxygen bridge forms.

Proteins condense between an amine group (NH_2) and an acid (COOH) group.

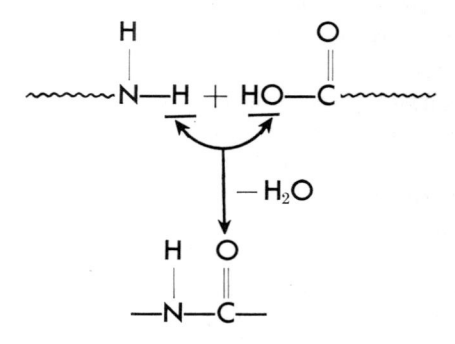

FIG. 33. *A peptide link*

It is easy to see how water splits out. The linkage has a special name: a "peptide" linkage.

It is interesting that cellulose, the strong fiber of plants like cotton, is much stronger than starch, although both are made of glucose units. Recall the nestling of the polyethylenes. The cellulose type of glucose does nestle; the starch type of glucose does not. That is the entire difference! When polymer chains can fit together, or nestle, the material becomes stronger.

Guncotton

Cellulose reminds me of another accident, which happened in 1846. A professor of chemistry in Switzerland was doing an experiment in his own kitchen. He was using a mixture of sulfuric and nitric acids. Suddenly, the glass container that held the hot mixture broke. Over his beautiful floor went the strong acids. The professor grabbed an apron and mopped up as fast as he could. The cotton apron was soaked with the acid mixture. He hung it over the stove. The apron dried, all right; but as soon as it dried, it exploded! Christian F. Schonbein (1799–1868), the professor, had made guncotton.

This is a way of putting "nitrates" on cellulose. After many later experiments by many chemists, methods were found of controlling the nitration process. Celluloid for movie film, rayon, photographic film, lacquers, and many other things were made from this accidental start.

Not all giant molecules have the same small molecule repeated many times. Many have two different molecules that repeat. Nylon is an example of this.

The easiest way to show how nylon is made is to illustrate the formulas of the two compounds that are used. It is easy to make nylon, and you can do it yourself. There are instructions in a number of commonly used books.

Many molecules have the potential of becoming monomers. Long chains that result have nestling possibilities, and can be very strong. For the manufacture of nylon, a variation of adipic acid is used, but the actual reaction is very similar to the one illustrated.

How Small Molecules Grow into Giants

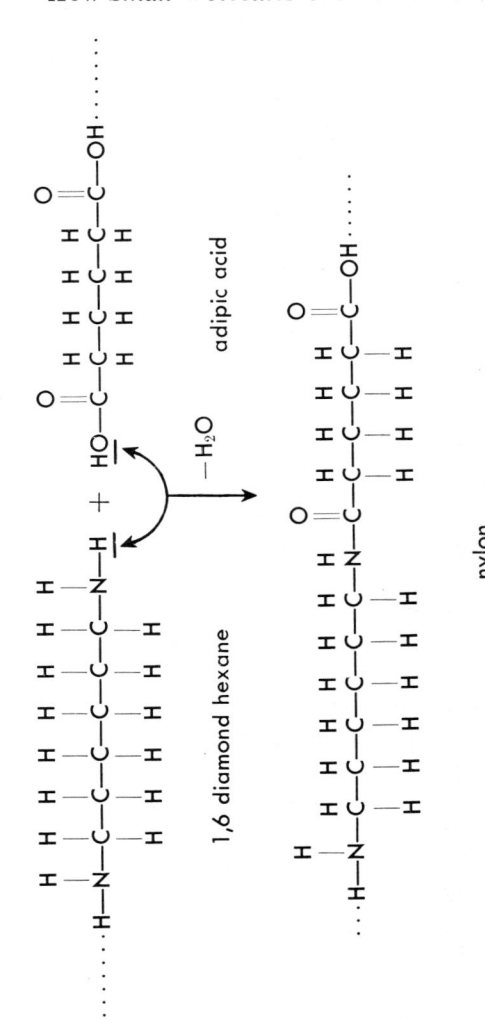

FIG. 34. *Structure of nylon*

VIII

The Power Locked in the Nucleus

In the vast dimension of time, it is only a small second that man knows, and a still smaller fraction of a second in which man has conceived real knowledge. What is the atom? It is the alphabet of all things. In that sense, without an understanding of the atom the language of the universe cannot be understood. What exactly is it that man has discovered about the atom? What is he capable of seeing in his mind's eye?

He sees a small but very heavy sphere or oval in the center and a great deal of space all around. In the space, crazy satellites dash in and out and around, in odd curves and in all directions. All things spin; the nucleus spins, the fast satellites spin. They move like whirling dervishes; they weave up and down while gliding like a figure skater doing tricks. If we dreamed such a picture, we might wake up in the middle of the night screaming.

What speed the satellites have! They go so fast that one cannot even judge where they are. Because of their speed, they look like clouds. What is the nature of the strange attractive force the nucleus has for these "electrons"? Through all of that unbelievable amount of space, it still holds the family together.

Then, try to imagine protons, all with the same charge, yet all huddled in the same tight bundle called a nucleus. You cannot conceive of such a thing unless you also imagine the neutrons. The protons must somehow have neutrons to keep them together. Let's examine the very smallest element of all, hydrogen.

Deuterium

Hydrogen has three isotopes. Number one is just a proton, with no neutrons at all. This isotope makes up more than 99 percent of all hydrogen atoms. The second isotope has, along with its single proton, one neutron. It is called deuterium and is famous for making the hydrogen bomb. There is only one deuterium atom in 7,000 atoms of hydrogen. The third and last isotope is one that, with the one proton, has two neutrons. It is called tritium and is very rare. There is one tritium atom in six million atoms of hydrogen. Tritium is in a special state. It keeps sending out pieces from the nucleus in order to change itself into another and more stable kind of atom. Whenever the nucleus is in such a state that particles in it cannot stay together for a long time, or do not fit properly together, it starts to lose parts of itself. It will continue to send out pieces until it gets into a stable condition. The sending out of pieces from the nucleus is called "radioactivity."

I remember when, as president of a chemistry teachers' club, I invited Dr. Harold C. Urey to speak before the group. Dr. Urey had discovered deuterium in 1931. I cannot now recall his exact words, but, in effect, he said:

"Deuterium was a missing link to the nuclear physicist. The physicist suspected its presence because the best calculations set the atomic mass of hydrogen at greater than it should be. True, the number came out a little bit more, not much at all, but it bothered them that it was not exact. It seemed to me that the best way to tackle the problem was to take into consideration two ideas. First, if an isotope of hydrogen did exist, it must weigh at least twice as much as the common hydrogen. Second, there wasn't much of it. Then I figured that a heavier liquid will not evaporate as quickly as a lighter liquid. So I made liquid hydrogen. I had to go down close to absolute zero to do so, but I made four liters of liquid hydrogen. Then, using good insulation, I evaporated the hydrogen as slowly as I could. I allowed it to evaporate until I had only one milliliter left. One milliliter is about twenty drops. This last amount of hydrogen we examined with a spectrometer. We found what we were looking for."

In a spectrometer, light is passed into the atoms of the material being tested. The electrons of the atoms pick up the energy of the light and become excited. They now have more energy than before. They soon fall back into their old state, emitting the energy that excited them. This emitted energy can be seen, very often, as colored lines on a photographic plate. Each element produces lines according to its own electron arrangement, so all are different. There is even a very slight difference in the lines of isotopes.

The hydrogen bomb is made of deuterium atoms. A recent hydrogen bomb was exploded (February 1976) underground in Nevada. The explosion was felt in Las Vegas, 150 miles away. The sun does the same thing, many times over. If we aim the spectrometer at the sun, we can distinguish elements like deuterium and tritium and helium. Deuterium atoms combine to make helium.

$$^2_1H + ^2_1H \rightarrow ^4_2He$$

This is called a "fusion" reaction, because two elements fuse to become one. But why is so much energy released?

Calculating Mass

In the helium nucleus there are two protons and two neutrons. The actual weight of a proton is about 10^{-24} gram. A number like that is just too hard to use all the time. Scientists labeled this mass a "unit mass" or "one." As science became more and more exact, the unit mass was defined with greater care. Today, we use the carbon 12 isotope and say that $1/12$ of its mass is the atomic unit mass.

According to this new scale, the mass of a proton is 1.0073. The mass of a neutron is 1.0087.

Now we can calculate the mass of helium:

2 protons	2 neutrons
1.0073	1.0087
× 2	× 2
2.0146	2.0174

$$\frac{\begin{array}{r}2\text{ protons} + 2\text{ neutrons}\\ 2.0146\\ +2.0174\end{array}}{4.0320} = \text{sum of the parts of the nucleus.}$$

According to all of our finest experiments, however, the actual mass of helium is only 4.003. The sum of the masses of the parts of the nucleus is greater than the mass of the nucleus itself? It doesn't seem to make sense. It makes sense if matter can be changed into energy. Albert Einstein said that $E = mc^2$. Energy equals the square of the speed of light multiplied by the mass. Square of the speed of light! What a number! Even a small mass change would yield a tremendous amount of energy. Imagine, in your mind's eye, a piece of matter expanding so much and so fast that it becomes energy—like light. Well, that's what happened to the rest of the helium nucleus. That is the secret of the hydrogen bomb and the heat of the sun. The loss in mass is called the "mass defect." The energy that is emitted is called the "binding energy." In other words, in order to break the nucleus into the same pieces that made it in the first place, you would have to put back the same amount of energy that came out; therefore the name, "binding energy."

This world we were born into—can you imagine how it itself was born? Small elements, like hydrogen and helium, combining and recombining, all with tremendous release of energy. It must have been like a sun! For millions of years heat constantly was produced; and then followed a long period of cooling. I wonder if anyone from another planet landed at that time and got his feet burned. This creature might have reported back to home base, "Barren. Hopeless. Fantastic heat. This must be what they call 'hell'."

After cooling, what remained? Which nuclei were well made, and which were made so badly that they were already beginning to decompose?

Stability of Elements

The small elements like helium, boron, nitrogen, oxygen, and even calcium mostly have the same number of protons as neutrons.

Such an arrangement seems to be very good in the smaller elements. The elements are stable. Then, in the larger nucleus, more neutrons than protons seem to be necessary. Up to bismuth (83 protons), the possibility of having stable isotopes still exists. But the next element, polonium, and all the elements that follow are constantly in trouble. They all throw out pieces; they are all radioactive. Remember that pieces will be thrown out until a stable arrangement of protons and neutrons is obtained. Peace and harmony must exist. An element may throw out a neutron and become a stable isotope. It may throw out protons, or various combinations. It would probably try to get into the below-84 group.

What happens to the particles that are thrown out? How do they affect us? One of my best friends was a researcher in the field of nuclear physics. He died in his thirties from radioactivity. Particles from unstable nuclei, traveling at great speeds, broke his body molecules. Radioactivity can be very dangerous.

It seems that *even* numbers of protons are a better combination for stability than *odd* numbers. Look at the famous six. Six elements make up almost all of the solid world: iron, silicon, sulfur, nickel, oxygen, and magnesium. The numbers of protons in the nuclei are respectively 26, 14, 16, 28, 8, and 12. All are even numbers. Some elements with odd numbers of protons have no stable isotopes at all. All of the atoms are constantly emitting pieces from their nuclei, changing, always changing, into something else. Element 43, technetium, is one. Promethium, with 61 protons, is another. Odd numbers of protons do not seem to make for stability.

What is the effect of the number of neutrons? Looking back at the chief elements of the solid world, their most common isotopes are:

       iron— 30 neutrons
     silicon— 14 neutrons
     nickel— 30 neutrons
    oxygen— 8 neutrons
      sulfur— 16 neutrons
magnesium— 12 neutrons

Again, all have an even number. The best arrangement seems to be even-even. Whenever there are elements that have odd numbers of protons, they seem to be stabilized by even numbers of neutrons. For example, the element fluorine has 9 protons. The most stable isotope of fluorine has 10 neutrons. The element sodium, with 11 protons, has 12 neutrons in its most common isotope. The larger element gold has an odd number of protons, 79; its most stable isotope has an even number of neutrons, 118. These, along with some others, are "odd-even."

Remember that all research in nuclear physics must take such statistical information into its plan of attack. It is vital for these physicists to know the kind of nucleus that is stable, and the one that is not.

We come, finally, to the last type of proton-neutron ratio. We come to the possibility of an odd number of protons and an odd number of neutrons. This "odd-odd" type is rare. The uncommon isotope of hydrogen, deuterium, has a nucleus of this type; it has one proton, and one neutron. The element boron has a stable isotope with five protons and five neutrons. Finally, the last one, nitrogen, has seven and seven for most of its atoms. Notice that all three are small elements. Small elements have a better chance of being stable. In general, the "odd-odd" type is rather rare among the earth's elements. This brings to mind a verse I composed:

1.
Look, saddle me even with even,
I do sit down right square,
All parts are here, not leavin'
I do fit a steady mare.

2.
But, saddle me not with odd stirrup,
And not with a one-eyed roan,
I'd sure fly out in the stir-up,
And break my odd-odd bone.

Radioactivity is an attempt by the nucleus to get itself into a stable position. The tendency seems to be to get below element number 84. For this, protons have to be emitted. Protons do not usually come out as single protons, but are ejected in the alpha particle. The alpha particle does quite a job. It removes from the clumsy nucleus a whole unit; a unit of two protons and two neutrons.

Getting to be a stable element of a lower atomic number can be a long and difficult task. One of the factors that is also important is the ratio of numbers. There is always an excess of neutrons over protons in the large nuclei. This excess must not grow too large, because then the nucleus becomes very unbalanced. The idea, then, is to eject exactly the right kind of pieces.

What does the radioactive nucleus eject? According to nuclear physics, there are several particles, but two are the most common. One of them is the alpha particle that we have talked about. It is the helium nucleus, two protons and two neutrons. The other is called a "beta" particle, and its charge and mass are the same as for an electron. The electron actually has a mass of 0.00054 on the atomic scale, but the number is so small that the mass is most often referred to as zero.

Along with particles ejected, an energy ray called the "gamma ray" comes out. It represents mass in the sense that mass has changed into energy. A positron is sometimes ejected. A positron has exactly the same mass as an electron and exactly the same amount of charge. The only thing different is that the charge is opposite. It is a plus electron. The positron is the thing that makes the proton plus. If a positron comes out, the proton becomes a neutron.

A very fine example of the conversions in radioactivity is the changing of uranium to lead.

Uranium is the 92nd element, and its most common isotope has a mass of 238. Uranium has a huge nucleus, but its actual radioactivity, the sending out of particles, is very slow. It sends out an alpha. The other elements that form could never exist if it were not for the radioactivity of uranium. These other elements, on the way down from uranium to lead, are much faster to decompose than their parent element uranium.

Not all of the isotopes of lead are stable. Uranium is element number 92; lead is element number 82. This means that uranium must lose 10 protons to get down to lead. It could eject five alpha particles:

It would become lead 218. Lead 218 would be so unstable that it could not exist. Perhaps the most stable isotope of lead is lead 206. Lead 207 and 208 are also stable. The ratio of neutrons to pro-

The Power Locked in the Nucleus

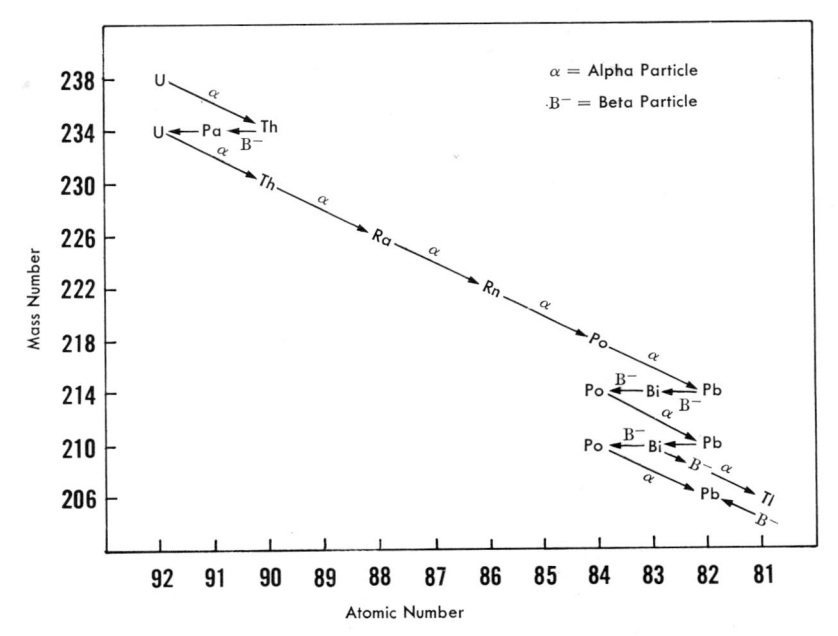

FIG. 35. *Changing uranium to lead*

$$^{238}_{92}\text{U} \xrightarrow{?} 5\,^{4}_{2}\text{He} + ^{218}_{82}\text{Pb}$$

FIG. 36a.

tons in lead 206 is $124/82 = 1.512$. This is a stable ratio, less than 1.55. When there are too many neutrons the nucleus is out of balance. The neutron/proton ratio is very important in the nuclear scheme. For example, we know that uranium 238 is radioactive. What is its n/p ratio? Its number of neutrons is 146, and its number of protons is 92. The ratio would be $146/92 = 1.587$. Such a ratio is unstable.

In the case of lead 218, the ratio would be $136/82 = 1.659$, really an impossible case. This is why uranium goes a different route to become lead. The first two steps are the emission of an alpha particle and then a beta particle:

1 $^{238}_{92}U \longrightarrow ^{4}_{2}He + ^{234}_{90}Th$

alpha

2 $^{234}_{90}Th \longrightarrow ^{0}_{-1}e + ^{234}_{91}Pa$

Beta

FIG. 36b.

The beta particle is an electron coming out of a neutron. The neutron has a neutral charge because it has both a plus proton and a minus electron. When it loses the electron it converts into a proton.

$$^{1}_{0}n \rightarrow ^{1}_{1}H + ^{0}_{-1}e$$

Neutron   Proton   Electron

Every time this happens, the neutron/proton ratio gets lower. It is therefore a very important step for stability.

If the series begins with the uranium isotope 235, the final result is lead 207. Of the two uranium isotopes, uranium 238 is by far the most common. Of every thousand atoms, 993 are of the 238 variety.

Now, if a neutron is sent into uranium 238, it simply becomes a 239 isotope. If a neutron is sent into uranium 235, it forms uranium 236. There is no isotope of uranium with a mass of 236. It simply explodes, breaking into two other elements and at least two neutrons. At the same time there is a mass conversion into energy according to the Einstein equation:

$$E = m \times c^2$$

energy = mass × *the velocity of light* × *the velocity of light*

The amount of energy is very great, as we have seen in the atom bomb. All that is necessary is that enough atoms of uranium 235 be put together to catch the ever-increasing flow of neutrons.

The scientists who discovered the terrible power of the atom bomb were aware of the possibilities of a "chain" reaction. In a chain reaction, one neutron would produce two, two would produce four, four would make eight, and so on. In fact, the buildup would be extremely fast, and the instantaneous conversion of mass into energy by countless numbers of uranium atoms would create a holocaust. Many of the scientists who worked on the atom bomb project were so dismayed at the possible demon that was being made that they left the project in protest. Today we know that one source of energy lies in the same power of the atomic nucleus.

IX

Let's Do Some Molecular Research

Some people like to quiz the teacher. I was writing the formula for glucose on the blackboard when a young lady raised her hand and said, "Sir" (I always feel older when someone calls me sir.), "How do you know that the $-OH$'s are placed exactly the way you have them? And how do you know that there are five of them?"

"Miss," I said, right back, "I refer you to last night's assignment. Didn't you do it?"

"I did the assignment, but I don't know how you know!"

I backed up defensively behind my big laboratory table. She had just fired a bazooka.

"You mean," I stammered, "you don't trust the book?"

"Well," she welled, "books aren't always right, are they?"

Then I did it. I didn't mean to do it. I guess I must have been afraid that she would finally accuse me of gross errors. "Then do it yourself," I blurted. I, a chess player, had put myself into checkmate. Her move was obvious.

The next morning, an hour early, she came rushing in. I was there, because I had exam papers to grade. She bustled around, gathering chemicals for the great experiments. After about fifteen minutes of this frantic work, she approached me and said,

"Where do I start?"

"Why, you start at the beginning." I gently shoved my unmarked papers aside. I knew what was coming.

"How?" She stood there, breathing gently, expecting wonders.

"Well, first you get yourself a research notebook. In this notebook the pages are numbered, and there is a large margin on the left. You date it on the first page when you start, and keep dating as you go along." I cleared my throat and continued.

"Then you have to find a good research library. A main public library in the city, or a college library. You're going to have to look up the original work on the structural proof of the glucose molecule."

"But sir," she sirred, "that means that I'll just be repeating."

"Yes," I said. "Repeat, and repeat again, if necessary. As you do so, keep your eyes and ears wide open."

"Well," she welled again, "eyes, O.K., but why ears? Is the stuff explosive?"

"No, it isn't explosive. But you must have friends who will try to help. Sometimes, another person will see something that you will miss. Don't be afraid to listen and take advice. There are lots of keen eyes and even keener minds around. Never be afraid to discuss a problem with a friend. Sometimes a researcher attacks a problem as a lioness attacks a gazelle. The lioness gets a good stranglehold and kills the little creature. The research worker, unlike the hungry lioness, needs the beautiful gazelle alive, and he must use a suitable approach. Seriously, many scientists have worked for years on a problem without decent results. Then someone new comes along and views the problem from a different angle. He solves it. Heartbreaking for the loser." I must have shed a tear.

"Get thee to a library," I continued, "and ask for the *Chemical Abstracts*. One of the volumes will have title references. Look up "glucose" and "structural proof." Get the names of the original publications and the journals. Locate the journals; read the original papers. If you have difficulty with them, get help."

"Sir," she asked politely, "could I come to you for help?"

"Of course," I said, wondering if she really meant to do all that work.

"Then," I continued, "write all of the reactions; write them in exact sequence, with numbers or letters. Most important, make sure to include the conditions. What are 'conditions'? The temperature, the pressure, how to add reagents—whether slowly or quickly, whether to add excess of a particular reagent. Regarding the pressure, many reactions end with the need to separate one product

July 12, 1963
Remarks

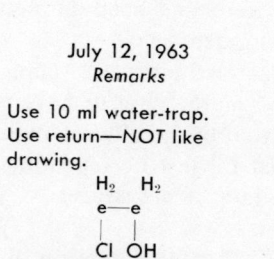

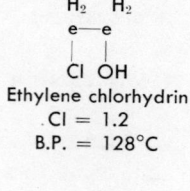

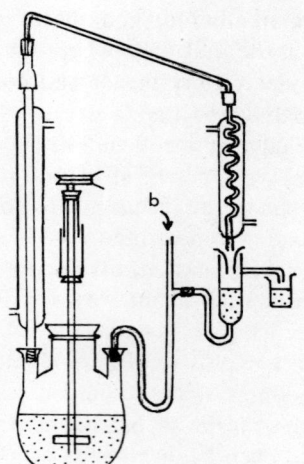

Use 10 ml water-trap.
Use return—*NOT* like drawing.

$$\begin{array}{cc} H_2 & H_2 \\ e\!-\!e \\ | & | \\ Cl & OH \end{array}$$

Ethylene chlorhydrin
Cl = 1.2
B.P. = 128°C

Done

1. Synthesis of the 1,2 diacetate ex diol
2. Synthesis of the 1,3 diacetate ex diol
3. Synthesis of the 1,3 chloro acetate ex CeICl$_3$ + diacetate
4. Synthesis of the 1,2 chloro acetate ex CeICl$_3$ + diacetate.

For comparison with # 3

5. Synthesis of chloroacetate (1,3) ex chloro hydrin.

I.R. gave very close similarity (no groups different) between #3 and #5.

Also work was done with ethylene diacetate + CeICl$_3$ to obtain

$$\begin{array}{cc} CH_2\!-\!CH_2 \\ | & | \\ Cl & OHc. \end{array}$$

Trouble

In all cases a peak at 2.8–3.1 was obtained on the I.R. This seems to demonstrate "free" OH. Perhaps water, or incomplete reaction, or hydrolysed— ??).

The problem of obtaining freedom from OH had to be settled before any further steps could be taken.

FIG. 37. *Entries in a research notebook*

from another. Distillation is often the answer. Some of the products may have to be distilled under low pressure. The low pressure will allow distillation at a lower temperature. The conditions of an experiment are a vital factor.

"Copy everything down very carefully into your research book. You could foul everything up if you made an error in copying. For example, suppose you found that, following every direction most carefully, the reaction did not work. The failure could be due to carelessness in copying. If, on the other hand, your research book is accurate, you may have done something very important in showing that the reaction actually does not work.

"Write balanced equations. In repeating equations from a book do not take for granted that they are perfectly balanced. Unbalanced equations can lead to wrong conclusions. Use the big left margin for several important details. First, physical constants. Every important compound placed in your book should have a listing in the left margin. Its boiling point, density, appearance, freezing point, and solubility in various solvents (water, ether, alcohol, benzene, chloroform) are important physical constants. Remarks like 'careful —carcinogenic' and 'trouble ahead' and 'ether—*no fires*' really do belong in that left margin. The left margin becomes more and more important.

"Finally, draw pictures. If there is a description of an apparatus, draw the picture and write the description. If there is only a reference to a reaction, and the actual reaction with its conditions is not given, you must go to a book on organic synthesis in order to get exact details. By this time you really want to get into the laboratory and get started. You seem to be wasting your time. Everybody goes through this phase of hating library work. But until you have had the experience of working in the laboratory without adequate preparation and accomplishing less than nothing, you will never know how important library work is.

"The reason for drawing an apparatus exactly as described in the original paper is to do a research within a research. You can try to invent a better way of obtaining the reaction. Making a better apparatus for a particular type of reaction has always given much pleasure. I can remember many times when just standing back and admiring a new and clever twist to the equipment gave me so much happiness. Sometimes a student will blow glass, make different and better shapes, invent new cooling systems, and use his imagination in ways that make the reaction better."

I stopped talking, looked at my watch, and prepared for my next class. My research student left for the library.

A whole week passed; the young scientist came to me with a problem. She had borrowed a couple of books from the library, and she carried them under her arm. Her right hand clutched a research notebook.

"Sir, there's something I don't understand."

"First," I said, "let me see your notes."

She put her books down and handed me the notebook. She stood still and anxious as I opened the hard-covered notebook to the first page. It was dated, it was neat, and it had titles. One page was headed, "Oxidation of the aldehyde." She had physical constants for every important compound. She had recorded boiling point, density, solubility, and appearance. She had even inserted some questions in the left margin. "Can't this be done with copper oxide?" and "What about side reactions?" and one question that said "Huh?" Not bad.

"The questions are good, but study the answers carefully. Take the copper oxide one, for example. I presume you are talking about the aldehyde oxidation. The trouble is that you are doing research work, and not just a little lab-book experiment. You should not mention just one chemical because you happen to have learned about it in class. You should list a number of possibilities, all you can find in the literature. You can scratch the bad ones off later. You must play the field. In your research, you must know more than the teacher and more than your textbook."

"But, sir," she interrupted, "I haven't got enough room in my notebook for all that data."

"I'll pretend I didn't hear that. I happen to have a couple of research notebooks in the stock room. Blank pages, ready to be filled. Now, what is it that you don't understand?"

"If I distill under low pressure, how do I know the new boiling point?"

"To know the new boiling point, you must first have a pressure gauge. This measures the exact pressure the substance is exposed to. Then you go to mathematics. There is a formula that converts boiling points at one pressure to boiling points at another. In all research, other branches of science are called upon for help. Mathematics, physics, biology. That is why a truly great scientist must have knowledge of many things.

"Now we come to labels. Each chemical that you use from the stock room has a label. The usual information on it is its name, its boiling point, its purity, and its molecular mass. Very often you will have to make solutions and put them into a glass container. You must be careful with the label. First, the name of the chemical; then the concentration. You must note the date the chemical was prepared, because it may spoil with time. Then, your initials.

"If certain reactions refuse to go well, it may be due to impure reagents. You may then have to test the purity of stock chemicals. You may have to go to spectrometers. There is one called an 'infrared' spectrometer, and one called an 'ultraviolet.' There are more, but you will probably find that the 'I.R.' or 'U.V.' are enough. If you find that the stock chemical is not pure enough, you may even have to go about making it more perfect. Many research chemists have spent long hours and days just purifying a 'pure' chemical."

While I was talking, class time had begun. I did notice, however, that the entire class was aware of my lecture on research. I felt that it might be a good idea to make a particular point. Addressing the entire class, I said:

"Not so long ago, a student asked me a question that caused the whole class to laugh. To quote him, he said,

'Why is the valence of magnesium always two plus?'

"The students all laughed because they had been told in very certain ways that the valences of all IIA elements are always two plus. It seemed to them that such a question was like doubting if two and two made four. But, you see, that's the difference between just doing your work and really thinking about it. With a quick idea in my mind, I decided to get the young man interested in research. The question was beautiful.

"Why do you ask?" I started my chess move.

"Well, I thought that the second electron would be harder to take off . . ."

"Aside from all of the foolish laughter, your question is a fantastic one. You show a great deal of insight. I enjoy hearing questions like that; it awakens my belief in the genius of mankind. When an atom that has two electrons in the valence layer loses one, it holds on more tightly to the second. It thus requires a lot of additional energy to kick off the second electron. The amount of energy needed to remove this second unpaired electron is the whole point we are trying to make. It amounts to twice and three times the energy used for number one. The question now is, why does number two always go?

"Magnesium compounds are made in one of two ways: first by burning, and second by reactions in water solution. All of the reactions are very exothermic—a great deal of heat is given out. This means that there is always enough heat around to release the second electron from the magnesium atom. Therefore, we always see magnesium salts as two plus magnesium. I will not answer the question as to whether it really can be one plus, but instead I will pose a research question and leave the rest to your imagination. If salts of magnesium one plus did exist, they would certainly be very useful. Now, here is the big question. Can you effect conditions such that the magnesium element would get just so much energy and no more; enough to release one electron, but not enough to release both?"

The classroom was so quiet that one could almost hear the thoughts of the people in the room. I remembered my own high-school days.

"When I was your age, we called column VIIIA the 'inert gas' column. We were taught with very great certainty that the 'perfect' elements of that column could never react with other elements to form compounds. Unwavering law; put anything else on a test and you get ten points off. After all, didn't the column VIIIA elements have all *paired* electrons? The elements helium, neon, argon, krypton, xenon, and radon had no chemistry. It turned out, however, that some people were not afraid of words like 'always' and 'never.' Just as it did here, with a young man's question. Thoughts ran in this way: as the atom, going down a column, gets larger and larger, the valence electrons get farther and farther removed from the attracting nucleus. We have mentioned this before, haven't we? Yet, having mentioned it, how many of you thought of it in connection with the noble gases? In any case, the valence electrons can be made to combine with a very fine grabber of electrons, like fluorine. It would be easier with the big ones, and much more difficult with the small elements like helium and neon.

"The brave people who dare to be different! The brave people who ask questions even though they may be laughed at! I can remember the reports of laughter when the automobile was invented; 'Get a horse!' was the favorite expression. The airplane made people howl with mirth. Please remember that."

I paused and asked for questions.

X

How to Win the Nobel Prize in Chemistry

No one knows the inner feelings and hopes of another human being. One knows only his own yearnings and desires to do something great. But sometimes we can guess. When Newlands, for example, was laughed at when presenting his paper for the first time, he must have felt despair and degradation. As we read the stories of people who finally gained supreme recognition by winning a Nobel Prize, we recognize the passions and perspiration that ruled their lives. These people worked hard. Sleep was necessary, but a waste of time. Often they had to be reminded that they had not eaten. Their passions were governed by success in the laboratory. There are many examples we could give, but we will use one to start.

Georg von Hevesy

In 1943 the Nobel Prize in chemistry was given to Georg von Hevesy for his work on the use of isotopes as tracers in the study of chemical processes. It all started in 1913, when Hevesy was working with Ernest Rutherford in the Institute of Physics at the University of Manchester. Rutherford was very anxious to obtain a supply of radioactive radium. The Austrian government owned a mine in Czechoslovakia that was rich in radioactive radium, and the government very nicely presented Rutherford with several hundred kilograms of this radium ore. There was only one problem. The ore also had lead in it. The lead kept absorbing the radiations given out by the radium. One day Rutherford met Hevesy in the basement where the stuff was kept. Rutherford said to the young scientist, "My boy, if you are worth your salt, you try to separate radium D from all that lead."

Hevesy was very enthusiastic. The great man had called upon him to do a very important piece of work. He could not fail! But try as he might, he could not separate the two. He worked hard for two years before he admitted defeat. He had failed Rutherford, and his heart was broken. He was in despair and considered the two years spent in futile attempt a complete waste.

But Hevesy examined his research notebooks with great care and came up with one conclusion. Radium D combined with the lead, and then the two could not be separated. Now he began to twist his project into a new one. Instead of attempting the impossible, he would take advantage of the fact that a radioactive substance had an inseparable attraction for lead. Radium D could be obtained, free from contamination, from other sources. It occurred to Hevesy that he could detect the amount of lead in a solution if he also put some radium D into the same solution. There was an instrument called an "electroscope" that registered radiations from atoms.

What Hevesy actually did was to "label" lead ions. He used a soluble lead salt like lead nitrate ($Pb(NO_3)_2$), in which the lead ions were freed when the salt was dissolved in water. He then added radium D. As we have seen, radium and lead combine and are inseparable. Another substance, thorium B, also radioactive, was used in the same way as radium D. Then Hevesy did a reaction with the lead ions. By adding a soluble chromate ($CrO_4^=$) salt, he made lead chromate ($PbCrO_4$). Lead chromate is practically insoluble; it stays as a solid on the bottom of the beaker. Now, in these insoluble salts there is always a tiny bit that does get into the water as ions. Up to this time, although it was known that a tiny bit dissolved, it was not known how much actually went into solution. After filtering away the solid, Hevesy evaporated the water and obtained a tiny bit of lead chromate. It was so small that the quantity could never have been calculated. But Hevesy, using the electroscope, obtained a radiation reading from the radium or thorium associated with the lead. He could then calculate how much lead chromate dissolved in the water.

There was much more to the work of Hevesy that finally gained him the Nobel award. From his beginning, he and others were able to make radioactive "tracers" that could detect many different things. Do metal atoms "stay put" in the metal, or do they move around? Is there an effect of temperature on atoms moving around in the solid phase? Research continues after one thing is discovered. More and more possibilities are opened for new discoveries.

Hermann Staudinger

We have discussed the making of very large molecules by the addition of many small ones. In 1953 Hermann Staudinger received a Nobel Prize in chemistry for his work in the field of macromolecular chemistry. Dr. Staudinger has written many books on chemistry, and it may be your good fortune to use one or more of them in your chemistry studies. The Nobel award that Staudinger won was for work started in the 1920's. In those years few scientists thought it possible that atoms could attach together to form very long chains. Staudinger maintained that as many as 10,000 to 100,000 atoms could join to form one big molecule. He pointed out that sometimes a few atoms would join together and then the ends would meet, forming a ring structure. This was well known. Then, he said further, if the ends do not join, it becomes possible for more and more atoms to tag on. A great long chain could then be formed. Since we know this so well today, it is amazing to us that this statement by Staudinger caused great controversy. For ten years the scientist could not really prove what he was saying. He found it practically impossible, with the methods of that time, to take the molecular weight of molecules that he claimed were "macromolecules." He was frustrated by his inability to prove his theory. It was not until the 1930's that his theory was recognized by everyone.

His theory had to be recognized; for by then the race to make bigger and better plastics was on. Bakelite was being produced. Nylon, acetate polymers, rubber polymers, vinyl polymers, and so on were being manufactured. Staudinger pointed out that nature had its own long, complicated molecules. Proteins, starches, nucleic acid polymers made up the living organism. He and his laboratory aides studied all of the macromolecules they could get their hands on. They studied their shapes and sizes, and figured out new methods for determining molecular weight. Through them, the new science of "macromolecular chemistry" was born. It is interesting that Staudinger's wife did a lot of work with him and published a number of brilliant papers on the subject. She was especially interested in the relationship between macromolecules and biology.

Textbooks are marvelous instruments for study, but one must not stop with the textbook. If Staudinger had taken the textbooks of the day at their word that molecular bonding could maintain just a few atoms, he would never have said: "There are molecules so large we should be able to see them with a good microscope." What a statement! The scientists of the day must have thought him absolutely crazy.

By all means, study the textbook with all your might. Get what you can out of it. Make it a basis for your research work. But do not consider it the last word on the subject.

Linus Pauling

In 1954 the Nobel Prize was awarded to an American. Linus Carl Pauling was born in Portland, Oregon, on February 28, 1901. His father was a druggist, and it is possible that Linus became interested in chemistry because he saw so many chemicals around. Pauling is seen often in public, giving lectures and acting in films where he discusses the nature of various bonds. He sometimes appears on television. He is a great talker, and one can listen to him for a long time. His interests are tremendously varied. The chemist thinks of him as a chemist. The biochemist thinks of him as a biochemist. The biophysicist thinks of him as a biophysicist.

Dr. Pauling won the Nobel Prize in chemistry for his "research into the nature of the chemical bond and its application to the elucidation of the structure of complex substances." Since 1919 his interests had lain in the field of how atoms bonded together. He did a great deal of work on the structure of crystals and how the ions arranged themselves in the crystal lattice (see the NaCl illustration in Chapter IV). He worked on the metallic bond and on the nature of magnetism. He worked on biological molecules; on the structure of antibodies and of blood cells; on diseases like anemia; on why anaesthetics put you to sleep. His work was, and is, so wide in its scope.

Linus Pauling wrote several books. Perhaps the most important is *The Nature of the Chemical Bond*. His book *General Chemistry* has been translated into at least nine languages and is used in colleges throughout the world. To give you an idea of how much work Pauling did, by the year he won the Nobel Prize, he had published about 350 papers in various fields. Many of the ideas concerning the way bonds form, the strengths of bonds, and the ionic nature of bonds can be traced to Pauling's work. He is the one who established a set of numbers called the electronegativity series, which put the elements into the order of power in attracting electrons. As he put it in his book, the "Heat of Formation" could be predicted by the difference in electronegativities. In other words, the strength of the bond itself could be predicted.

Vincent du Vigneaud

Now, I would like to do a little tracing, starting at the beginning and illustrating the eventual winning of a Nobel award. Let's take the case of the award of 1955. In that year it was given to Vincent du Vigneaud for his work on "biochemically important sulfur compounds, especially for the first synthesis of a polypeptide hormone."

Vigneaud's first teacher in biochemistry was Professor H. B. Lewis. Professor Lewis was extremely enthusiastic about sulfur. The young student was infected with this enthusiasm and started to think "sulfur." Then we see Vigneaud working at the same University of Illinois with Dr. C. S. Marvel in organic chemistry. As he worked, he became convinced that the structure of the molecule had a direct bearing on how it would work in a living organism. Our story continues with his work, molecules containing sulfur; a concentration on how to detect sulfur, how to isolate it.

Then, one day, Vigneaud heard a lecture by W. C. Rose, a professor of biochemistry at Illinois. The lecture was a very exciting one about the discovery of insulin. Vigneaud was stimulated to investigate the new compound. He found it has sulfur in it. As you probably know, insulin is the hormone that performs a number of important functions in the animal body. One of the functions of which we are most aware is the action of affecting the membrane of body cells in such a way as to allow glucose to enter the cell. In cases where the insulin does not function, or is insufficient, the organism has a disease called diabetes mellitus. Since so many cases of diabetes were known, insulin was, and is, an extremely important compound.

The next step finds Vigneaud working with Professor J. R. Murlin at the University of Rochester Medical School. The subject? The chemistry of insulin. He is still involved with the secret of sulfur. He finds sulfur compounds in insulin. He finds the important sulfur–amino acid, cystine. He isolates pure cystine from insulin. He and a number of graduate students start working with cystine to see if it would form bonds with other amino acids. Amino acids are the little building blocks of the large protein molecules.

So the chain continues; more and more links are formed. Vigneaud never actually discovered the exact insulin molecule. That was left for some years later. But on the way, he made a few other important molecules called hormones. Hormones are also made of amino acids, but they are smaller, having fewer amino acid units than do the proteins. Vigneaud and the laboratory made the three-amino-acid unit glutathione, which is present in all living things and has many functions. It functions as a "co-enzyme" in many reactions, and seems to be important in maintaining the red blood cells.

Vigneaud was still on the trail of the sulfur compounds. Cystine became his most important lead. From the back area of the pituitary gland, he isolated a hormone called "oxytocin." Oxytocin is the hormone that helps the contraction of smooth muscles and the ejection of milk from the mammary glands of females. Counting cystine as two amino acids (two cysteine make one cystine), oxytocin has eight amino acids. After a tremendous amount of work and sleepless nights, Vigneaud succeeded in actually making the molecule of oxytocin. This was the beginning of all the great work on proteins that is still going on today.

As you see, from a little beginning, the interest in sulfur, a keen mind went on and on until great things were accomplished. I must emphasize, with all kinds of research in mind, that sheer persistence is often more important than brilliant deduction. You know, as I do, students who are brilliant but don't do much work. They don't care; their high grades come too easily. On the other hand, to do research at my side, I want someone who can study a problem from many angles, and work, and work (and keep a good notebook).

Walther H. Nernst

Two people, wise in chemistry, look at the equation:

$$H_2 + Cl_2 = 2\ HCl$$

One says, "Hydrogen gas combines with chlorine gas to give hydrogen chloride gas. Nice, clean equation, easy to balance."

The other chemist, who happens to be the German Walther H. Nernst (1864–1941), looks at the reaction and says, "Hmm, the reaction is explosive. It goes very, very fast. How can two molecules, each with perfect pairs, react in such haste?"

A question has been asked. A question must be answered.

Nernst then said, "Light breaks some of the chlorine molecules into their atoms."

$$Cl_2 \quad = \quad :\ddot{C}l\cdot \ + \ :\ddot{C}l\cdot$$
Molecule   Atom   Atom

"The chlorine atom is very active, having that unpaired electron ready to form a partnership. If the chlorine atom hits a hydrogen molecule, it will combine:

$$:\ddot{C}l\cdot + H_2 = HCl + H\cdot$$

"The reaction of combination to produce hydrogen chloride plus hydrogen is based on the fact that the hydrogen-to-chlorine bond is a much stronger bond than the hydrogen-hydrogen bond. The heats of formation will tell you that; the electronegativity values will tell you that, also."

Then, in my imagination, Nernst continues his lecture.

"The hydrogen atom formed is ready and able to join, its unpaired electron sticking out like a sore toe. If it hits a chlorine molecule,

$$H\cdot + Cl_2 = HCl + \ :\ddot{C}l\cdot$$

it will immediately attach, forming a stable H — Cl bond, and, at the same time, throw out an active chlorine atom."

If a strong light like a magnesium flare were placed near a glass tube containing a mixture of hydrogen and chlorine gas, the entire tube would probably shatter with the explosion. The formation of free atoms would make such a quick reaction understandable. In 1956, Nikolai N. Semenov won the Nobel award for experimenting with the way reactions work. He studied the formation of particles with an unpaired electron. These particles with the single electron are called "free radicals." Semenov worked with organic molecules, illustrating reasons why many reactions work through the process of free radical formation, whereas some do not. For anyone familiar with the carbon compounds that make up organic chemistry, Semenov's ideas are very clear and understandable. They are the extension of the Nernst theory of the reaction between hydrogen and chlorine gases.

Many research experimenters use something already discovered, repeat it, and extend it. They even win Nobel Prizes with it. Recall the instructions given to the young woman in the previous chapter, "Repeat, and repeat; and keep your eyes and ears wide open." If I were to try to influence you to go after the Nobel Prize in chemistry for the United States, I would not say:

"Student, keep in mind the great award ahead of you; always remember . . ."

Nonsense! I would say, instead:

"Student, love your work. Love it by knowing it, really knowing it. See pictures in your mind; work hard at understanding, not necessarily at passing exams. That comes with understanding. The pleasures of understanding are intense and beautiful. The rewards of research are in the doing and in the finding. Should other awards follow, the soul will be flattered and honored. But the joy has been there, and is there, and is forever."

APPENDIX I

Glossary

"A" columns: The "A" columns in the Periodic Table contain the elements with unique chemical similarities. Each of these elements has the same number of electrons in the outer shell.

acid: A substance that gives up protons to another substance. This is the Bronsted-Lowry definition.

A substance that has an empty orbital in its outside shell. For example, H^+, BF_3, $AlCl_3$. This is the G. N. Lewis definition.

actinide series: A number of elements, following radium in the seventh period, in which an inner level, 5f, is being filled with electrons. The elements are called "rare earth metals."

active elements: Elements that lose electrons easily are active metals. Elements that gain electrons readily are active nonmetals.

addition: A reaction in which unsaturated organic compounds add atoms to become more saturated.

A reaction in which two or more substances combine to form one.

addition polymer: A polymer formed by addition reactions between unsaturated compounds called monomers.

air resistance: The resistance of air molecules to the passage of beams or rays.

alchemist: One who practiced the study of alchemy from the 13th to the 17th centuries. He concerned himself mostly with three ideas: (1) the transmutation of other metals into gold; (2) the finding of a universal remedy for disease; and (3) the discovery of a universal solvent.

alcohol: An organic compound containing a hydrocarbon group and one or more hydroxyl (-OH) groups.

aliphatic compound: An organic compound that consists of straight or branched chains.

alkali metals: The Group IA metals are called alkali metals. The hydroxides of these metals are called "alkalis."

alkane: A compound in the saturated series of hydrocarbons (C_nH_{2n+2}).

alkene: A compound in the unsaturated series of hydrocarbons. It is characterized by having a double bond between two carbons (C_nH_{2n}).

alkyne: A compound in the unsaturated series of hydrocarbons that is characterized by having three bonds between two carbons (C_nH_{2n-2}).

alpha particles: Nuclei of helium atoms. The nucleus contains two protons and two neutrons. The particles are emitted at high speeds from certain radioactive elements called "alpha emitters."

amine: A derivative of ammonia (NH_3), in which one or more of the hydrogen atoms are replaced by organic groups; for example, CH_3NH_2 (methyl amine).

amino acid: An organic compound that contains both the organic acid group carboxyl (-COOH), and the organic base amino (-NH_2). Both are commonly on the same carbon; for example, the amino acid glycine,

$$\begin{array}{c} H \\ | \\ H-C-COOH. \\ | \\ NH_2 \end{array}$$

analysis: The science of identifying a chemical substance, usually done by breaking it down to simpler substances.

anode: The electrode at which electrons are deposited. It is often called the "plus" electrode. Oxidation occurs at this electrode.

antibodies: Proteins produced in the blood as a reaction to the presence of antigens. Antigens are usually poisons to the body and are neutralized by the antibodies.

"B" elements: Elements, beginning with element number 21, scandium, in which an inside energy level is being filled with electrons.

These elements are also called "transition elements" and have different characteristics from the *A* column elements.

calces: The substance of a mineral after it has been heated or treated with an acid. It is a very old term, used by alchemists, and it really refers mostly to metallic oxides.

caoutchouc: A thick liquid from the hevea tree; it becomes a soft, flexible rubber on standing.

carbon tetrahedron: When a carbon atom is bonded to four atoms or groups, its four single bonds are directed to the corners of a regular tetrahedron.

carbonyl group: The organic group in which carbon is attached to oxygen by a double bond.

carboxyl group: The organic acid group, made up of carbon, two oxygens, and a hydrogen. It is usually written -COOH.

cathode: The electrode at which electrons are gained. It is also called the "minus" electrode. Reduction occurs at the cathode.

cathode rays: Rays or streams of electrons going from the cathode to the anode in a discharge tube.

chain reaction: A series of reactions in which the reactants produce more reactants.

chair form: The puckering of a cyclohexane-type structure into a chairlike form in order to relieve angular stress. The six-membered ring glucose is believed to exist in a chair form.

chemical property: The characteristics of a substance in the way that it combines with other substances.

compound: A substance composed of simpler substances in definite proportions.

conjugate system: A system in which double bonds alternate with single bonds in an organic compound.

conservation of mass: The theory that states that, in a chemical reaction, mass can be neither gained nor lost.

coordinate covalent bond: A combination of two atoms to form a covalent bond. Only one of the two atoms bears the bonding pair of electrons; the other atom has an empty orbital.

covalent bond: The force that holds atoms together by a shared pair of electrons.

cyclic compounds: The term usually refers to organic compounds in which carbons and other atoms arrange themselves into cyclic structures. Some cycles are more stable than others. The five- and six-membered rings are found most often.

deuterium: An isotope of hydrogen known as "heavy hydrogen." Besides the one proton, deuterium has one neutron; it is, therefore, twice as heavy as hydrogen.

dipole: A molecule in which one end is somewhat positive and the other end somewhat negative.

double bond: A bond between two carbons (or other atoms) that consists of two pairs of electrons. A very common double bond is the one between carbon and oxygen (-C=O).

electrode: A conductor by which an electric current either enters or leaves an electric apparatus.

electron: A negatively charged particle, associated with the atom. It has a mass of 9.11×10^{-28} gram and an atomic mass unit value of 0.00055.

electronegativity: The property of an atom to attract the bonding pair of electrons. The higher the electronegativity, the greater the power to attract electrons.

electroscope: A device used to detect the presence of charged particles. Since radioactive substances bombard the air with charged particles and induce other charged particles to form, an electroscope can detect radioactivity.

element: A substance that cannot be further decomposed by ordinary chemical means.

endothermic reaction: A chemical change in which there is an absorption of heat energy.

energy level: A region around the nucleus of an atom where electrons with certain energy values probably travel.

ether: An organic oxide. The oxygen atom is situated between two carbon atoms: -C-O-C-.

excited state: When electrons of an atom are in a higher energy level than they normally occupy.

fission: The splitting of an atomic nucleus: some of the nuclear mass is converted into energy.

free radical: An atomic or molecular species that has an unpaired electron. Free radicals are intermediates in reactions. Examples of free radicals are H·, Cl·, ·CH$_3$.

fusion: A nuclear reaction in which small nuclei combine to form larger nuclei. The reaction is accompanied by the conversion of mass into energy.

gamma ray: A high-frequency, high-energy wave, most often associated with radioactivity.

gel: A colloidal dispersion of a liquid in a solid.

guncotton: A substance produced by treating cotton with a mixture of nitric and sulfuric acids. It explodes by percussion.

heat of formation: The heat either absorbed or given off in the formation of a molecule of a compound from its elements.

hormones: Substances synthesized by living organisms to act as physiological regulators.

hydrogen bond: A weak chemical bond formed between the hydrogen of one polar molecule and the electronegative element of another polar molecule. Hydrogen bonds may also occur within the same molecule.

induction effect: The electrostatic effect of one molecule on a neighboring molecule. A minus charge repels electrons on the neighboring molecule, and a plus charge attracts electrons.

inert: Unable to enter into any chemical changes.

inorganic: Compounds other than hydrocarbons and their derivatives.

insulin: A hormone produced in the pancreas. Insulin affects the organism's ability to utilize sugar.

ion: atom or group of atoms with an unbalanced electrostatic charge.

ionic bond: A type of bonding in which there is a transfer of one or more electrons from one atom to another.

isomers: Compounds with the same molecular formula, but with a different arrangement of atoms.

isotope: Atoms of the same element, but with different numbers of neutrons.

lattice: The pattern of points in space in a crystal, which describes the arrangement of atoms. It is also called a crystal lattice.

macromolecule: A very large molecule, made up of small molecules bonded together.

magnetic field: The charge around a magnet that can attract or repel charged particles, or polar molecules.

mass: The measure of the quantity of matter in a body. The measure of the inertia of a body.

mass defect: The difference between the mass of a nucleus of an atom and the sum of the parts that make up the nucleus. The parts should equal the whole, but they do not. The sum of the parts is greater. The difference is called "the mass defect."

mass number: The total number of protons and neutrons in the atomic nucleus.

metal: An element that loses electrons in a chemical reaction. It has luster and is a good conductor of heat and electricity.

metallic bond: A type of bonding in which electrons from the valence layer of the metal atoms form a unique energy level and surround the entire crystal lattice of the metal atoms.

molecule: The smallest part of a covalent substance that can exist free and still maintain the properties of the substance.

momentum: The product of mass times velocity.

monomer: The repeating unit in a polymer.

negative charge: The name given to the kind of charge associated with an electron. The area with an excess of electrons is said to have a negative charge.

noble elements: The gases of column VIIIA on the Periodic Table are called "noble gases." Since 1962 they are no longer considered to be inert. Compounds have been made of xenon, krypton, and radon.

nonmetals: Elements that normally gain electrons in chemical changes. They have poor luster and poor conduction of heat and electricity.

nonpolar molecule: A molecule in which the centers of positive and negative charges coincide; it has charge symmetry.

nucleus: The central part of an atom. It contains protons and neutrons.

neucleon: A particle of the nucleus; like protons and neutrons.

neutron: The neutral particle found in the nucleus of the atom; it has about the same mass as a proton.

nestling: The manner in which linear polymers may fit together to produce a stronger material.

octet: The outer shell of an atom that has its s and p orbitals filled with eight electrons.

orbital: The most probable space around a nucleus in which electrons with the requisite energy values may be found.

organic: Compounds of the element carbon; hydrocarbons and their derivatives.

oxide: A compound consisting of oxygen and (usually) one other element.

oxygen bridge: The connection between monomers in a polymer that consists of an oxygen atom. For example, starch is a polymer of glucose molecules connected by oxygen bridges.

peptide linkage: The linkage caused by joining the acid group (-COOH) of one molecule with the basic amino group (-NH$_2$) of another molecule:

$$\begin{array}{c}-\text{C}-\text{N}-\\ \parallel\ \ |\\ \text{O}\ \ \text{H}\end{array}$$

period: A horizontal row of elements in the Periodic Table.

Periodic Table: The table of all known elements arranged according to their atomic numbers and their chemical behavior.

phlogiston: A gas or fire thought to be part of a substance in the days of alchemy. Phlogiston could be gained or lost, thus changing the chemical nature of the substance.

plastic: A natural or synthetic material that can be shaped while soft into a desired form and then hardened by heating and cooling. It is most often a polymer.

polar covalent bond: A covalent bond in which one of the atoms attracts the shared pair of electrons more than the other. One part of the molecule is more negative than the other.

polymer: A large compound made of small repeated units; the units are called monomers.

polywater: A fictitious polymer in which the monomers were supposed to be water molecules.

polysaccharide: A polymer in which the monomers are monosaccharides (like glucose).

positive rays: A stream of particles that are positively charged, like protons or alpha particles.

positron: A particle equivalent in mass to the electron. Its charge is exactly opposite to the charge of an electron, and it is sometimes called a "positive electron."

protein: A polymer made by living organisms that consists of small molecules called amino acids; examples are silk, hair, skin, muscles, enzymes.

proton: The positively charged particle present in the nucleus. Its charge is equal and opposite to the charge on an electron. Its mass is 1,836 times greater than the mass of an electron.

quantum: The smallest unit of energy. It is expressed by a basic equation: $E = h\nu$, where E is a quantum (singular) h is Planck's constant, and ν is the frequency of vibration of the energy-bearing system.

quantitative analysis: The branch of chemistry that deals with the determination of the amount of each component of a chemical system.

quantum mechanics: The mathematics that seeks to determine the electronic configuration around an atom. It is also called "wave mechanics."

radioactivity: The process of decomposition of unstable nuclei; particles like the alpha and beta particles are emitted from the nucleus.

ratio method: A method commonly used to determine atomic masses. By means of chemical reactions, chemists first determine the ratio of elements in a compound. Then, by weighing the compound, they determine the individual weights of the component elements.

rayon: A name used for textile products produced by certain chemical reactions on native cellulose (cotton or wood). The final product is also cellulose, but in an altered form.

recipe: The word used in the days of alchemy to mean an equation or set of equations. There were some very strange recipes.

ring structure: Bonded atoms in a closed arrangement like a ring. The most stable rings are five- and six-membered rings. The atoms that make up the rings are mostly carbon, then nitrogen, oxygen, and sulfur.

salt: An ionic solid that consists of a positive ion and a negative ion. This is the G. N. Lewis definition. Another definition has salt as a metal ion and a nonmetal ion combined.

saturated compound: An organic compound in which the carbon atoms have only single bonds between them.

sextet of electrons: Six p electrons of the benzene ring that, instead of forming the second bond of the three double bonds, surround the carbon atoms of the ring. Thus, benzene does not have any discrete double bonds.

shear strength: The resistance of a material against being cut or sheared across.

silicones: Organic silicon compounds where the stable silicon-oxygen link forms giant polymers. Hydrocarbon groups are attached to the silicon atoms:

$$\begin{array}{c} R \\ | \\ -Si-O-Si-O-Si- \\ | \\ R \end{array} \begin{array}{c} R \\ | \\ \\ | \\ R \end{array} \begin{array}{c} R \\ | \\ \\ | \\ R \end{array}$$

spectrometer: An optical instrument used for producing and analyzing light spectra. It has either a glass prism or a diffraction grating and sends a specific wave length through the substance being analyzed.

starch: A polymer whose monomer is glucose. The glucose units are in the form of rings, and the rings are connected by oxygen bridges. The similar substance in animals is called "glycogen."

substance: A species of matter in which all parts have identical properties.

sugar: A member of a group of compounds called carbohydrates. Sugars have sweet tastes; they are either single units, called "monosaccharides," or double units called "disaccharides." Table sugar is a disaccharide consisting of glucose and fructose.

synthetic rubber: Man-made rubber, imitating natural rubber, which is based on a compound called polyisoprene. The monomer is isoprene:

$$CH_2=C-CH=CH_2$$
$$|$$
$$CH_3$$

Synthetic rubber has a number of variations.

Telluric Screw: A way of placing elements in a classification scheme, on a spiraling paper, set on a cylinder. The word "telluric" means the earth (referring to earth elements).

temperature: The average kinetic energy of the particles of a system.

tensile strength: The strength of a substance in its resistance to being pulled apart.

tetrahedral structure: A molecular structure in which atoms or groups of atoms are so spatially arranged as to form the corners of a tetrahedron.

theory: An explanation of observed happenings and relationships. The explanation is largely verified.

thermometer: An instrument used to measure temperature. It commonly employs the principle of expansion and contraction of mercury with changes in temperature.

transmutation: The changing of one element into another. This can be done by a change in the atomic number.

triad: A group of three elements having similar chemical properties. In 1828 Döbereiner set some elements into triad units.

triple bond: A type of bonding in organic chemistry that has three bonds (three pairs of electrons) between two carbons.

tritium: The hydrogen isotope that has a mass number of three. Its nucleus has one proton and two neutrons. It is the rarest of the three hydrogen isotopes.

unsaturated compound: Organic compounds that have double or triple bonds between carbon atoms.

vacuum tube: A tube from which the air molecules have been almost completely extracted. It is most often used for electrical purposes.

valence: The combining power of an element.

valence layer: In "A" elements, the outermost principal energy level of an atom. "B" elements may have more than one valence layer.

van der Waals forces: The electrostatic attraction between nonpolar molecules. The attraction is caused by the asymmetric distribution of the electrons of the molecules. This uneven distribution causes a slight polarity; this in turn, produces induced polarity in neighboring molecules. Weak intermolecular bonds can then form.

volume: The amount of three-dimensional space that is occupied.

wave: A means of transmitting energy.

wave frequency: The number of waves that pass a set point in one second.

wave length: The distance between one point on a wave and the corresponding point on the next wave.

X ray: A radiation of very short wave length and high wave frequency. It has good penetrating power and can go through liquids and many solids.

APPENDIX II

Books for Further Reading

Azimov, Isaac. *Building Blocks of the Universe.* New York: Abelard-Schuman, 1961.
The author discusses all of the elements known at the time of writing.

Chedd, Graham. *Half-way Elements.* Garden City: Doubleday & Co., 1969.
Boron, silicon, germanium, arsenic, antimony, and tellurium are discussed as to their role in technology.

Frisch, Otto R. *The Nature of Matter.* New York: E. P. Dutton & Co., 1973.
This book describes the atomic theory of matter and discusses the discovery of new particles of the nucleus.

Gamow, George. *The Atom and Its Nucleus.* Englewood Cliffs, N.J.: Prentice-Hall, Inc., 1961.

Hughes, Donald J. *The Neutron Story.* Garden City: Doubleday & Co., 1959.
The book discusses the history of the neutron discovery, the mesons within the neutron, neutron capture, and other details about the neutron.

Ley, Willy. *The Discovery of the Elements.* New York: Delacorte Press, 1968.
The author offers a chronological discussion of the discovery of 104 elements, beginning in antiquity.

Romer, Alfred. *The Restless Atom.* Anchor Books, 1960.
This book traces the development of knowledge about the X ray, radium, radioactivity, and the atom.

Spruch, Grace, Marmor, and Spruch, eds. *The Ubiquitous Atom.* New York: Charles Scribner's Sons, 1975.
This book consists of excerpts from the series of booklets "Understanding the Atom," prepared by the Atomic Energy Commission.